More Praise for

INTO *the* GRAY ZONE

"The model of how pop science involving sensational subjects should be done."

—*The Globe and Mail* (Toronto)

"Fascinating . . . With remarkable clarity, Owen punctuates his findings with concise dispatches on the human condition and the disparities between what is considered quality of life and what some consider an inhumane, dysfunctional existence. . . . A striking scientific journey that draws hopeful attention to how the brain reacts, restores, and perseveres despite grave injury."

—*Kirkus Reviews*

"[A] fascinating memoir . . . *The Gray Zone* reads like a thriller as [Owen] recounts his and his teams' efforts to explore [the] 'gray zone.' . . . Owen's enthusiasm for his science crackles from the pages. His determination to fight for the scores of voiceless gray-zone patients he encounters, to prove they're 'thinking, feeling people,' is hugely thought-provoking and deeply moving."

—*Mail on Sunday* (UK)

"Meshing memoir with scientific explication, Owen reveals how functional magnetic resonance imaging can probe the deep space of trapped minds. It's a riveting read, from the march of technology and tests for neural responses—such as imagining playing a game of tennis—to extraordinary personal accounts of the 'gray zone' by partially recovered pati

—*Nature*

"*Into the Gray Zone* weaves a fascinating tale using medical data, heart-wrenching case studies, and [Owen's] own personal experiences."

—*Good Housekeeping*

"Groundbreaking . . . a fascinating and accessible account of cutting-edge science, and of those whose lives have been altered in an instant. . . . Owen's enthusiasm for his subject is infectious. . . . This book will be required reading for anyone sitting by a loved one's bedside, caregivers, doctors, ethicists, lawyers, and philosophers."

—*The Sunday Times* (London)

"Vivid, emotional, and thought-provoking . . . Owen's story of horror and hope will long haunt readers."

—*Publishers Weekly*

"*Into the Gray Zone* is required reading for anyone who wants to explore the outer limits of consciousness, and the human spirit. Neuroscientist Adrian Owen takes us on a gripping, often harrowing journey into the most mysterious realm of human experience: the twilight zone between life and death."

—Joshua Horwitz, author of *War of the Whales: A True Story*, winner of the 2015 PEN/E. O. Wilson Literary Science Writing Award

"A fascinating and highly readable book, written with evangelical fervor . . . gripping and moving."

—*New Statesman* (UK)

"[A] remarkable book . . . Through examinations of human brains damaged by trauma, tumors, infections, and vascular accidents,

[Owen] attempts to explore the nature of consciousness. . . . [His] experiments have allowed vegetative-state patients with residual consciousness to connect with the external world."

—*Literary Review* (UK)

"Although he has written hundreds of research papers about his work, *Into the Gray Zone* is Owen's first book pulling it all together in fast-paced prose. Readers should prepare to be educated, yes. But more satisfyingly, they should prepare to be fascinated, astonished, and, at times, moved to tears."

—*Winnipeg Free Press*

"*Into the Gray Zone* is both a crystal-clear description of cutting-edge neuroscience from one of the pioneers in the field, and a set of intensely personal stories about patients in the twilight of consciousness. . . . One of the most moving and gripping science books you're ever likely to read."

—Daniel Bor, author of *The Ravenous Brain: How the New Science of Consciousness Explains Our Insatiable Search for Meaning*

"This is a great book—immensely moving, profound, and engaging, with a zest for life and science that bubbles off the page. . . . Adrian Owen has lived the dream of a neuroscientific discovery that changes thinking about a terrifying medical condition and how patients and their families can be given the answers they crave. Reading how this happened will make you alternately laugh, gasp, and cry."

—John Duncan, author of *How Intelligence Happens*

"An unforgettable book. Owen weaves together stories of human resilience in the face of extraordinary adversity with an account of his own groundbreaking research, and in so doing takes us on a deeply moving journey to the very frontiers of consciousness. I couldn't put it down."

—Tim Bayne, author of *The Unity of Consciousness*

"An amazing book that challenges basic assumptions about what it means to be a person! What's on display here is a curious branch of brain research that is both fascinating and, frankly, terrifying. . . . It should be required reading for anyone interested in the brain, and especially for all those who care for patients thought to be in a vegetative state."

—Katrina Firlik, author of *Another Day in the Frontal Lobe: A Brain Surgeon Exposes Life on the Inside*

"Provides fascinating insight into cutting-edge neuroscience and the power of the human psyche. . . . Time and again, we are taken to the edge of our seats, reflecting on what it means to be alive and how hope can triumph in the most tragic of circumstances."

—Richard Wiseman, bestselling author of *The Luck Factor* and *59 Seconds: Think a Little, Change a Lot*

"Truly moving and inspirational . . . an uplifting testament to the power of scientific curiosity and the extraordinary resilience of the human spirit. This book delivers an eloquent message: even in the most desperate circumstances, there can be hope."

—Roger Highfield, former editor of *New Scientist* and coauthor of *SuperCooperators* and *The Arrow of Time*

"Simply unputdownable . . . Taking my evening bath while dipping into the opening pages of *Into the Gray Zone*, I finished three hours later, with the water cold. What kept me in the bathtub is Owen's account of communicating with the most impaired neurological patients—those unfortunate individuals whose damaged bodies and brains often put them at a greater distance from us than an astronaut lost in space."

—Christof Koch, PhD, president and chief scientific officer, Allen Institute for Brain Science

"What an amazing read! . . . The book is a real page-turner, both because it unpacks the complexities of modern neuroscience in an accessible way and because it directly confronts profound ethical questions."

—Melvyn Goodale, PhD, coauthor of *Sight Unseen: An Exploration of Conscious and Unconscious Vision*

"Captivating . . . In this book, which will bring new hope to many, we see Owen explore new realms of consciousness—ones experienced by patients who are devastated by brain injury yet surprisingly endowed with thought, feeling, and memory."

—Kevin Nelson, author of *The Spiritual Doorway in the Brain: A Neurologist's Search for the God Experience*

INTO *the* GRAY ZONE

A NEUROSCIENTIST EXPLORES THE MYSTERIES OF THE BRAIN AND THE BORDER BETWEEN LIFE AND DEATH

≈

ADRIAN OWEN

SCRIBNER
New York London Toronto Sydney New Delhi

Scribner
An Imprint of Simon & Schuster, Inc.
1230 Avenue of the Americas
New York, NY 10020

First Scribner trade paperback edition June 2018

Interior design by Kyle Kabel

Manufactured in the United States of America

1 3 5 7 9 10 8 6 4 2

Library of Congress Cataloging-in-Publication Data is available.

ISBN 978-1-5011-3520-0
ISBN 978-1-5011-3521-7 (pbk)
ISBN 978-1-5011-3522-4 (ebook)

For Jackson

In case I'm not here to tell you the story myself

That you may see the meaning of within
It is being
It is being

—John Lennon and Paul McCartney

CONTENTS

INTO *the* GRAY ZONE

PROLOGUE

I'd been watching Amy for almost an hour when she finally moved. She had been sleeping when I arrived at her bedside in a small Canadian hospital a few miles from Niagara Falls. It seemed unnecessary, even a little rude, to wake her. I knew there was little point in trying to assess vegetative-state patients when they are half-asleep.

It wasn't much of a movement. Amy's eyes flicked open; her head came up off the pillow. She stayed that way, rigid and unblinking, her eyes roving around the ceiling. Her thick dark hair was cropped short, but perfectly styled, as though someone had been working on it only moments earlier. Was this sudden movement simply the result of automatic firing of the neural circuitry in her brain?

I peered into Amy's eyes. All I saw was emptiness. That same deep well of emptiness that I had seen countless times before in people who, like Amy, were thought to be "awake but unaware." Amy gave nothing back. She yawned. A big openmouthed yawn, followed by an almost mournful sigh as her head collapsed back onto the pillow.

Seven months after her accident, it was hard to imagine the person Amy must once have been—a smart college-varsity basketball player with everything to live for. She'd left a bar late one night with a group of friends. The boyfriend she'd walked out on earlier that evening was waiting. He shoved her and she toppled, slamming her head on a concrete curb. Another person might have walked away with a few stitches or a concussion, but Amy was not so lucky. Her brain hit the inside of her skull. It pulled from its moorings. Axons stretched and blood vessels tore as ripples of shock waves lacerated and bruised critical regions far from the point of impact. Now Amy had a feeding tube surgically inserted into her stomach that supplied her with essential fluids and nutrients. A catheter drained her urine. She had no control over her bowels, and she was in diapers.

Two male doctors breezed into the room. "What do you think?" said the more senior of the two, looking straight at me.

"I won't know unless we do the scans," I replied.

"Well, I'm not a betting man, but I'd say she's in a vegetative state!" He was upbeat, almost jovial.

I didn't respond.

The two doctors turned to Amy's parents, Bill and Agnes, who'd been patiently sitting while I observed her. A good-looking couple in their late forties, they were clearly exhausted. Agnes gripped Bill's hand as the doctors explained that Amy didn't understand speech or have memories, thoughts, or feelings, and that she couldn't feel pleasure or pain. They gently reminded Bill and Agnes that she would require round-the-clock care for as long as she lived. In the absence of an advanced directive stating otherwise, shouldn't they consider taking Amy off life support and allowing her to die? After all, isn't that what *she* would have wanted?

Amy's parents weren't ready to take that step and signed a consent form to allow me to put her in an fMRI scanner and search for signs that some part of the Amy they loved was still there. An ambulance shuttled Amy to Western University in London, Ontario, where I run a lab that specializes in the assessment of patients who have sustained acute brain injuries or suffer from the ravages of neurodegenerative diseases such as Alzheimer's and Parkinson's. Through incredible new scanning technology, we connect with these brains, visualizing their function and mapping their inner universe. In return, they reveal to us how we think and feel, the scaffolding of our consciousness, and the architecture of our sense of self—they illuminate the essence of what it means to be alive and human.

Five days later I walked back into Amy's room, where I found Bill and Agnes by her bedside. They looked up at me expectantly. I paused for a moment, took a deep breath, and then gave them the news that they hadn't allowed themselves to hope for:

"The scans have shown us that Amy is not in a vegetative state after all. In fact, she's aware of everything."

After five days of intensive investigation we had found that Amy was more than just alive—she was entirely conscious. She had heard every conversation, recognized every visitor, and listened intently to every decision being made on her behalf. Yet she had been unable to move a muscle to tell the world, "I'm still here. I'm not dead yet!"

≈

Into the Gray Zone is the story of how we figured out how to make contact with people such as Amy, and the profound effects for science, medicine, philosophy, and law of what has become a new and rapidly evolving field of inquiry. Perhaps most important, we have discovered that *15 to 20 percent* of people in the vegetative state who are assumed to have no more awareness than a head of broccoli are fully conscious, although they never respond to any form of external stimulation. They may open their eyes, grunt and groan, occasionally utter isolated words. Like zombies, they appear to live entirely in their own world, devoid of thoughts or feelings. Many really are as oblivious and incapable of thought as their doctors believe. But a sizable number are experiencing something quite different: intact minds adrift deep within damaged bodies and brains.

The vegetative state is one realm in the shadowlands of the gray zone. Coma is another. Comatose people do not open their eyes and look completely unaware. In the Disney version of *Sleeping Beauty* (which most parents know all too well), Aurora's condition resembles coma, akin to a bewitched slumber. In real life, the picture is far less romantic: disfiguring head injuries, contorted limbs, broken bones, and wasting illnesses are the norm.

Some people in the gray zone *can* signal that they're aware. Referred to as minimally conscious, they occasionally respond to requests to move a finger or track an object with their eyes. They seem to fade in and out of awareness, occasionally emerging from some deep pool of oblivion, breaking the surface and signaling their presence before sinking back into the murky depths.

Locked-in syndrome is not technically a gray-zone state, but it is close enough to give us insight into what life might be like for some of the people we scan. Locked-in people are fully conscious and can typically blink or move their eyes. Jean-Dominique Bauby, French editor of *Elle* magazine, was a famous example of someone locked in. A massive stroke left him permanently paralyzed except for the ability to blink his left eye. With the help of an assistant and a writing board, he composed *The Diving Bell and the Butterfly*, a memoir, which took two hundred thousand blinks to complete.

Bauby vividly recounted his experience: "My mind takes flight like a butterfly. There is so much to do. . . . You can visit the woman you love, slide down beside her and stroke her still-sleeping face. You can build castles in Spain, steal the Golden Fleece, discover Atlantis, realize your childhood dreams and adult ambitions." Of course, this is Bauby's "butterfly": the mind unbound, unconstrained by physicality or responsibility, free to flit here and there. But Bauby was also locked inside the "diving bell," an iron chamber from which there is no escape and which sinks ever deeper into the abyss.

Back at Amy's bedside a few days after her fMRI scans, I again sat watching her closely, desperately wanting to know what she was thinking and feeling. All of those convulsive movements and spasmodic gurgles. Was her experience like Bauby's? Had she entered Bauby's imaginative realm of freedom and possibility? Or was her inner world an excruciating prison from which there was no escape?

Following our scans, Amy's life changed beyond recognition. Agnes would barely leave her bedside, reading to her more or less constantly. Bill popped in each morning, delivering the daily papers and updating Amy on the latest family gossip. A constant stream of friends and relatives visited. Amy went home on weekends, and parties were held on her birthdays. She was taken to the movies. The care staff always introduced themselves to her, explaining that they were going to wash or change her before approaching her bedside. Every intervention, every drug, every change of routine, was carefully explained. After seven months in the gray zone, Amy became a person again.

I didn't delve into this new field of science with anything resembling a clear idea in mind of what I wanted to do. The beginning felt like a fluke, an offhand coincidence. Yet as I look back, it's clear that what set this story in motion points to the inner fabric that binds all of us together in ways that are monstrously complex and impossible to anticipate. My explorations into the gray zone emerged out of something dark and strange that happened in a leafy, genteel suburb of south London on a warm July day twenty years ago. . . .

CHAPTER ONE

THE GHOST THAT HAUNTS ME

≈

People don't live or die, people just float
She went with the man in the long black coat
—Bob Dylan

The scientific process works in mysterious ways.

As a young neuropsychologist at the University of Cambridge, studying the relationship between behavior and the brain, I fell in love with Maureen, a Scottish woman who was also a neuropsychologist. We met in the fall of 1988 in Newcastle upon Tyne, an English city sixty miles from the Scottish border. I had been sent up to Newcastle University to solidify a collaborative relationship between my boss, Trevor Robbins, and Maureen's boss, the improbably named Patrick Rabbitt, who was doing innovative work on how the brain ages. Maureen and I were thrust together. I was immediately charmed by her dry wit, amazing head of chestnut hair, and lovely eyes that would tightly close whenever she laughed, which she did all the time. I was soon returning to Newcastle upon Tyne for less academic reasons, driving six hours up and back through murderous weekend traffic in my ancient Ford Fiesta, a banged-up piece of junk that I'd picked up for £1,100 from my first paycheck.

Maureen introduced me to music. Not the bland early-eighties glam rockers in eyeliner, hair spray, and jumpsuits such as Adam and the Ants, Culture Club, and Simple Minds that I'd been infatuated with through my adolescence, but the music that I still carry with me. Passionate music that told stories about land and history mixed with relationships and burning desire. The driving, soulful Celtic-based music of the Waterboys, Christy Moore, and Dick Gaughan. Maureen's brother Phil, who lived in St. Albans, about forty-five miles from Cambridge, quickly persuaded me that a future without a guitar in hand was no future at all and took me to buy my first axe—a Yamaha that I still own and always will.

After some months of commuting between Cambridge and Newcastle upon Tyne, I moved sixty miles south to London because that's where the patients I was studying were being treated. I continued to work as a neuropsychologist, paid by my boss in Cambridge, and signed on for a PhD at the Institute of Psychiatry at the University of London, driving between the two cities several times a week to fulfill the obligations of both posts. It was a grueling schedule, but I loved the work. Maureen gave up her job in Newcastle, took a position in London, and we soon bought our own place—a small third-floor one-bedroom apartment that was a short walk from the Maudsley Hospital and the Institute of Psychiatry in South London, where we both were based.

As a building, or set of buildings, the institute is extremely disappointing—a sprawling jumble that lacks a physical presence to match its formidable academic reputation. My office was in a prefabricated building, or portacabin, as we call them in the UK. Freezing in winter, sweltering in summer, it shook each time the main door slammed. We were promised more permanent digs every year: the portacabins would be razed. But I would return decades later and discover, to my surprise and amusement, that there they were, probably still housing aspiring PhDs.

The initial flush of excitement and romance that Maureen and I felt about moving in together was soon replaced with the more humdrum business of driving to see patients all over southern England, sitting in endless lines of stationary London traffic, searching in vain for vacant parking spots within walking distance of our home, and jump-starting my Fiesta when it decided not to start in the morning—which was all the time.

Working at the institute and the Maudsley, it was impossible not to be moved by the patients: legions of depressives, schizophrenics, epileptics, and demented souls pacing the drafty corridors. Maureen, an empathic, caring person, was deeply affected by them. She soon decided to train as a psychiatric nurse. Despite the doubtless nobility of this calling, her decision struck me as an abnegation of what could have been a glittering academic career. She began spending long evenings out with her new colleagues while I stayed home, writing and rewriting my first scientific papers, describing the shifts in behavior of patients who had had pieces of their brains removed to alleviate epilepsy or eradicate aggressive tumors.

The histories and stories of what had happened to these patients once their brains had been tampered with fascinated me. One patient I worked with had minimal frontal-lobe damage but became wildly disinhibited as a result. Before his injury he was described as a "shy and intelligent young man." Postinjury he abused strangers in the street and carried a canister of paint with him to deface any public or private surface he could get his hands on. His speech was littered with expletives. His wild behavior escalated: he persuaded a friend to hold his ankles while he hung from the window of a speeding train, a lunatic activity by any measure. His skull and most of the front part of his cortex were crushed when he crashed headlong into a bridge. By some circular twist of fate, his minor frontal-lobe injury led directly to major damage to the same part of his brain.

Perhaps the most bizarre case I encountered concerned a young man with "automatisms"—brief unconscious behaviors during which you are unaware of your actions. Automatisms are typically caused by epileptic seizures that start in the temporal or frontal lobes and then quickly spread—an escalating cascade of neuronal firing that engulfs the entire brain. During these episodes, patients hang in a kind of gray zone. Their eyes remain open, and they are strangely animate and seemingly purposeful in their actions. These usually include routine activities: cooking, showering, or driving a familiar route. Following the episode, the patient regains consciousness and often feels disoriented but has no memory of the event.

My patient was a lanky youth with wild hair whom I tested for memory impairments following surgery that he had received to combat seizures. He was also the defendant in a murder trial. The victim was his own *mother*, strangled while she was securely locked in the house with her son. Just the two of them. The case turned on his being a martial arts expert with a history of epileptic automatisms, and he could (although the evidence remained entirely circumstantial) have killed her through a series of routine martial arts maneuvers and remained entirely unaware of this dreadful act.

When I assessed his memory using what were then our state-of-the-art computerized tests, I sat near the door—a strategy I had seen in numerous TV crime dramas. I didn't feel safe. I needed a weapon. All this now seems ludicrous, but there I was, sitting in a closed office with a man who was accused of killing his own mother with his bare hands without even knowing that he'd done it! If he *had* done it, could he be judged responsible? I wasn't sure. The thinking then and now was that automatisms, rather than expressing subconscious impulses, are automatic programs firing in the brain, completely outside our control. If he had been a carpenter, he would have been sawing a piece of wood rather than karate-chopping his mom.

Could his brain make him kill again? That was the uppermost question in my mind. What could I use to defend myself? The office around me was stacked high with papers, books, and the paraphernalia of scientific investigation—not exactly an armory. Beside the desk I spied a squash racquet. I clutched it, mulling over some vague plan to parry the young man's blows. Fortunately for both of us the session passed without incident. I have often thought what an odd sight it would have been: the patient attacking me like a ninja while I tried to swat him about the head with a squash racquet.

The work was enthralling, but all the while I was losing touch with Maureen. Within a year of buying our apartment, the relationship fell apart. We were going in different directions: me into a career in science and her into a job in psychiatric care. Something had changed between us. I couldn't understand why she'd lost the sense of shared wonder about the brain and how it is affected by damage and disease. I couldn't understand the appeal of what felt like simply caring for a problem rather than trying to solve it. I'd made the decision, some years earlier, not to pursue a traditional medical career. I'd never wanted to be a physician, listening to people's ailments and dishing out medication according to standard protocols. I wanted to try to *understand* the mysteries of the way our minds work and perhaps discover new approaches to treatment and cures. That's what neuroscientists do. I thought that I had my eye on the bigger picture, but I was probably just insufferably self-righteous, driven by the ambition and idealism of a young scientist. I thought we might be able to understand and then *cure* Parkinson's and Alzheimer's diseases.

I was also dazzled by what then impressed me in my naïveté as the glamour that a high-flying career in neuroscience might offer.

My boss was sending me to exotic locales to give talks in his stead. At an academic conference in Phoenix, Arizona, I found myself in a hot tub in the desert with two other English neuroscientists. Can you imagine? The day before we had all been plodding through the perpetual precipitation and dreariness of England, and then there we were, luxuriating among the cacti.

I must have been a bit smug when I came home from these trips. Maureen and I had a running argument about the rights and wrongs of psychiatric care, science for science's sake, and the innate tensions between scientific discovery and medical care.

"It's all very well studying these people," I remember Maureen saying. "But helping them deal with their problems is a much better use of resources."

"If we don't do the science, these problems will persist!" I countered.

"Science might help someone down the line, years from now. But it mostly comes to nothing. And it doesn't help patients who donate their time to *your* research projects, naively assuming that you are going to make their lives better."

"I *do* tell them that my research is not going to help them personally."

"Wow. Aren't you *nice*?"

Our running argument had undertones of England versus Scotland. Since the beginning of time, the Scots have felt exploited by the English, whom they see as cold, bloodless mercenaries while they are passionate, earthly, and honest. In retrospect, our care-versus-pure-science positions echoed this age-old conflict.

Eventually, I met someone else and I left Maureen, moving out in 1990 just as the UK economy and housing market collapsed. Our £60,000 apartment was suddenly worth £30,000. We had an enormous negative equity. The interest rate on our mortgage doubled, which was barely manageable while Maureen lived in the apartment. Things rapidly deteriorated when she also moved in with someone else. To make the mortgage payments we were forced to rent the apartment to Brazilian friends, but Maureen wanted nothing more to do with it. I collected rent, paid the mortgage, and took care of taxes and repairs. Maureen and I were no longer on speaking terms—just sending angry letters back and forth. I ended up sleeping on the floor of a friend's apartment in North London, a whole hour's drive through rush-hour traffic to see my patients at the Maudsley Hospital. The previous owners had taken their cats but left the fleas. It was a miserable time.

That same year, as I went from patient to patient in South London documenting their brain injuries and their stories, strange things started to happen to my own mother's health. She began experiencing blinding headaches and behaved in odd ways. One afternoon she disappeared for several hours and upon her return explained that she had been to see a film at the local theater. She hadn't been to the movies in years and certainly not on her own in the middle of the day. She had just turned fifty, and our family doctor concluded that her menopause was to blame, both for her headaches and curious, unusual excursions. He couldn't have been more wrong. One evening at home as she watched TV with my father, it became more clear that something was seriously amiss.

"What do you think of the woman's dress?" my father asked, referring to a woman on the far left side of the screen.

"What woman?" My mother couldn't see the woman. In fact, she couldn't see anything in her left visual field at all.

Whatever was causing her headaches and odd behavior was now also affecting her vision. Simple tasks, such as crossing the street, became too dangerous for her to tackle alone. Imagine that you are no longer able to see anything in one part of your visual field (what you see from left to right as you look straight ahead). The problem is that our brains are remarkably good at adapting to change, and in situations such as this, they can literally reconfigure our worldview to what can be seen, completely ignoring what can't. The missing part does not appear as empty space or as blackness, as one might imagine—it ceases to appear at all. Crossing the road with no awareness of anything on her left side was no longer something that we were going to let my mother attempt alone.

A CT scan revealed that my mother had an oligoastrocytoma growing inside her brain—a cancerous tumor that was pushing its way into the folds of her cortex, interfering with her behavior, affecting her moods, changing how she saw the world, and altering her whole sense of being. We were all devastated. Suddenly, my family's life and my chosen career were colliding in the most diabolical way imaginable. If she'd been sent for surgery and lost part of her brain as a result, my mother could easily have ended up as a patient in one of my research studies. It was a nightmarish thought.

I was now on the other side of the fence. No longer the detached young scientist but a distraught family member—a situation I'd seen many times among the patients and families that I had been visiting in and around South London. Unfortunately, unlike the tumors in many of those patients, my mother's was deemed inoperable, and she began round after round of chemotherapy, radiotherapy, and steroid treatment. Swelling around a brain tumor puts pressure on surrounding tissue—that's what causes the headaches. Steroids reduce the swelling and relieve those symptoms. My mother's hair fell out and she became bloated (a frequent side effect of steroids).

Fortunately for my family, my sister had qualified as a nurse in 1990 and had been working at the Royal Marsden Hospital, a famous London institution that is dedicated to cancer diagnosis, treatment, research, and education. She gave up work in July of

1992 to care for my mother at our family home. That same month I submitted my PhD thesis, which told the stories of patients with brain disorders, including tumors similar to the one my mother was battling. Before I could formally graduate, I had to defend my thesis, and that would take some months to arrange. By then it was clear that my mother would soon die. I desperately wanted her to see me graduate with a PhD. I called the main administrative office at the University of London and explained the circumstances. Without hesitation they agreed to let me "graduate" despite my not yet having competed the full requirements of the PhD—that would come later. We never told my mother. She was at my graduation, although she may not have been aware of what was going on. I vividly remember my father and I hauling her out of her wheelchair into one of the seats in the auditorium, me dressed in my flowing graduation gown, her in the best clothes we could find that still fit her. We lost our grip and she fell helplessly into the aisle. These are the consequences of progressive brain damage that no one tells you about. In between what you once were and what you eventually become is a grueling adaptation to the deterioration of your day-to-day abilities as tasks become increasingly difficult and finally impossible.

Soon after graduation day, my mother slipped into her own gray zone, not quite there, but not quite gone. Still living at home, now bedridden in the ground-floor dining room since she could no longer climb stairs, she slipped in and out of consciousness from the massive doses of painkillers and sedatives administered by our family doctor. Sometimes she recognized us, sometimes she didn't. Sometimes she was lucid, sometimes she made no sense at all. My brother flew home from the States, where he was in the throes of his own postdoctoral studies at NASA's Goddard Space Flight Center in Maryland, and we spent the last few days together as a family. She died in the early hours of the morning on November 15, 1992. We were all at her bedside when she finally stopped breathing.

Many dark days followed, but in a strange way something good came of my mother's death. After four years of meeting those affected by brain damage and documenting their lives, I got to be on the other side and experience what it is like to watch someone you love get slowly drawn into the abyss. Whether that experience made me even more determined to pursue a career in brain research I do not know, but it certainly prepared me for the many encounters I would have with brain-injured patients and their families in the years to come. I knew firsthand what they were going through, and I felt for them. I wanted to help in any way I could.

Shortly before my mother's death, I had been offered a postdoctoral position in Montreal, Canada, and now I jumped at the chance to move abroad. I was more than ready to walk away from the ruinous apartment and failed relationship with Maureen and my mother's death from a brain tumor at fifty. I was through with England and took a three-year position at the Montreal Neurological Institute.

≈

Arriving at "the Neuro" at the end of 1992 to work with Michael Petrides, then the head of the Department of Cognitive Neuroscience, was a significant slice of good fortune. Michael was passionate about brain anatomy and always keen to embrace any new approach or method that might help illuminate how the brain does mental activities such as memory, attention, and planning. Over the next three years, we spent many hours poring over his drawings of the frontal lobes, scribbling little notes about what each area of the brain probably did and designing new tests that would show us how different parts of the brain contributed to memory. I would go away and program them on my IBM 386—state-of-the-art then but woefully underpowered by today's standards.

This was the year that what were called positron-emission tomography (PET) "activation studies" took off, driven, in part, by developments in the computing industry that allowed us to capture large data sets and digital images of the brain in action. From the launch of the Hubble Space Telescope and the Human Genome Project, computers were revolutionizing every aspect of science. And we were part of that revolution.

Volunteers for PET activation studies would lie in the scanner and be injected with small amounts of a radioactive tracer, and then we'd ask them to perform a task: remembering an unfamiliar face we flashed in front of them, for example. The principle was delightfully simple: those parts of the brain that were working hardest required more oxygen, which was delivered in the blood. Blood flow increased to areas involved in a task. We could literally *map* the movement of blood around the brain with our PET scanner.

It was a neuropsychologist's dream come true. No longer would we have to wait for a special patient to come through the door with damage to one specific part of the brain in order to deduce what that brain area did. Now we could simply put healthy people in the scanner and ask them to perform our cognitive tests while we watched their brains spring to life and reach the very same conclusions.

Much of the early work was confirmatory, but that just added to the excitement. For instance, we'd known for some years that the fusiform gyrus, an area on the undersurface of the brain, is involved in face recognition; patients with damage to that area have problems recognizing people they know, a condition known as prosopagnosia, or "face blindness." But to see the ultimate confirmation of this, when this area lit up in a group of healthy participants as they looked at a series of familiar faces presented on the computer screen, was astounding.

≈

We naively thought we were going to be able to quickly unlock all the secrets of the brain, PET scan by PET scan; but we soon ran into the limitations of what we had at first thought was limitless technology. First among them was the so-called radiation burden. For each scan we gave participants a safe but significant dose of radioactivity. This limited the number of scans we could give any one person, which seriously restricted how many scientific questions we could ask in any one study.

The second problem with PET was that the changes in blood flow that we detected were so small that it was virtually impossible to identify them with a single scan. We had to repeat scans to build a clear picture of what was happening in the brain. We inevitably hit the radiation burden, sometimes before we'd answered a single scientific question to our satisfaction. The answer was to average the data from multiple participants. Indeed, the signals from the brain were so small that this is what we had to do most of the time.

That posed a third problem—our scientific conclusions were not about individuals but groups. Rarely could we say what a particular part of the brain was doing in any one person. Rather, our conclusions would typically take the form of "On average, across the group . . ."

A fourth limitation of PET was timing. A single scan took between sixty and ninety seconds, and what you saw at the end was the sum total of everything that happened during that period. Individual "events" slipped under the radar. Imagine a task where we asked participants to view and remember a series of faces during a ninety-second scan. It was hard to know whether the brain activity that we'd see after the analysis was complete was caused simply by the seeing of the faces, by the remembering of those faces, by some of the faces and not others . . . the list of unknowns went on and on. In spite of all of these limitations, those of us who studied the brain thought all our Christmases had come at once. From the minute I set foot in the door and began designing PET activation studies I was hooked.

One of my early successes showed that one area of the frontal lobes was crucial for organizing our memories. It wasn't the place where memories were stored or the part of the brain that committed information to memory. Rather it dictated "how" memory should be organized. Visualize trying to remember where you parked your car this morning in a lot you use each day. How do you remember today's parking spot and not confuse it with the place where you parked yesterday, or the day before, or last week? You could use a landmark, such as a tree or a nearby building, but you've probably used all those landmarks before and you are bound to get confused by them. You have to make a special kind of memory *decision*—you have to decide that, of all the parking spaces that you have in your memory from days gone by, *this* is the space that you are going to remember today. You have to label this particular space as special and especially relevant *for today*. This process is an example of what we call *working memory*, which is a special kind of memory that we only need to retain for a limited period, until the information is used, in this case until you successfully retrieve your car at the end of the day. Then the whole process starts again the next day.

Your working memory chips in whether it's a telephone number remembered just long enough to punch it into your phone, the face of the stranger in the crowded room remembered just long enough to return the pen she lent you, or the parking space that you picked this morning for your car. No one knows what happens to these ephemeral memories. Do they just vanish into thin air? Evidence suggests that they seem to be "overwritten" by subsequent working memories. We seem to have a limited capacity for this type of brain function, which, when exceeded, leads to the inevitable removal of one memory in favor of another.

These types of studies dovetailed neatly into other areas. We started to scan patients with Parkinson's disease to try to understand why it is that they, in particular, have problems with working memory. Unlike Alzheimer's patients, if you show patients with Parkinson's disease a picture that they have never before seen, they will have little trouble recognizing it later. But show them a whole series of pictures and ask them to remember one or two in particular, and the task becomes much harder. Why? It's similar to the parking-space problem. Their problem is not with laying down memories, but with *organizing* them in such a way that retrieval is possible in the face of fierce competition.

≈

During my three years in Montreal I kept the London flat afloat. Maureen and I hardly communicated. Our occasional conversations were terse, clipped, and filled with frustration on both sides. Then, in 1995, my former Cambridge boss, Trevor Robbins, called. A new brain-imaging facility—the Wolfson Brain Imaging Centre—was being set up at Cambridge's Addenbrooke's Hospital, and they needed someone with my expertise. As a research fellow in the Department of Psychiatry, I would run the first brain-activation studies at Cambridge, supervise students, and start to put together a lab of my own. They had a PET scanner, and Trevor convinced me that if I got my foot in the door, it could lead to a more permanent position at Cambridge. No permanent positions were on the horizon in Montreal.

So I went home to the UK in 1996. Much had changed in England since I had left; in particular, brain scanning had taken over. If you weren't scanning brains, you were nothing, and the UK was leading the pack. What hadn't changed was my strained relationship with Maureen. We both found it too painful to see each other and avoided meeting up at all costs. It had been four years since our breakup, and whenever I thought of our apartment and failed relationship, I felt frustrated and confused. How could we have ever been so in love and wanted to build a life together? And how had all that changed? What could possibly have been going on in her head? It made no sense. She was an absolute enigma.

Then, one July morning in 1996, a colleague called. Maureen had been found unconscious, lying beside her bike on a steep hill near the Maudsley Hospital. It was initially assumed that she'd crashed into a tree and knocked herself out cold. But it turned out to be worse—much worse. Tests revealed that she had suffered a subarachnoid hemorrhage, a ruptured brain aneurysm; a weak area in the wall of an artery had released blood into her skull. Aneurysms can be caused by a multitude of factors: family history, gender (they're more common in women), high blood pressure, and smoking.

Yet again my personal life and my professional life collided in the most abysmal way imaginable. I had assessed many patients who were recovering from the effects of a subarachnoid hemorrhage just like Maureen's. Many of them had problems with memory, concentration, and planning—the hemorrhage and the surgery that was necessary to treat it affected their lives forever, disrupting their thoughts, impacting their memories, and altering their personalities unpredictably. Just like my mother, Maureen could have ended up in one of my own research studies! Unfortunately, Maureen's aneurysm wreaked even more havoc than was usual for most of my patients, and she was quickly diagnosed as being in a vegetative state—I was told that she would not likely survive. Although it was probably not the first time I had heard the expression *vegetative state*, it was certainly the first time it registered.

Imagine my shock. What had happened to Maureen? What did being in a vegetative state mean? Was she dead or alive? Did she know where or who she was? She was gone, but she wasn't. How could she still be living and breathing, waking and sleeping, and yet be somehow so completely absent? This was made far more confusing by my feelings for her. How does it feel when someone you have been so close to, and then so far away from, is suddenly rendered vegetative? It feels very strange indeed.

With proper care, vegetative patients can live a long time. Several months after her brain injury Maureen was flown back to Scotland to be closer to her parents. She was kept alive, seemingly oblivious, by the people and the machines that helped feed and hydrate her. To prevent bedsores, she was regularly turned by the nursing staff. They bathed her with warm sponges, washed her hair and clipped her nails. They changed her bedding and her clothes. They talked to her, bright and chipper in the morning. ("And how are we today, Maureen?") On weekends, they dressed her and she was moved by wheelchair to her parents' house, where members of her loving family would often visit her.

It did not *consciously* occur to me that perhaps some form of consciousness could still reside in the brain activity of people such as Maureen, who were outwardly completely nonresponsive. Yet maybe that seed of an idea, outlandish as it seemed at the time, was planted. Perhaps it was a trigger. A calling to do something more useful with the experience I had acquired in using these incredible new technologies to lay bare the workings of the brain—something that Maureen would have endorsed. She had been so passionate that science should not be "science for science's sake": it should actually *help* people. Perhaps this was a chance for me to do just that.

CHAPTER TWO

FIRST CONTACT

≈

> I can listen no longer in silence. I must speak to you by such means as are within my reach.
>
> —Jane Austen

Enter Kate. Age: twenty-six. Occupation: nursery-school teacher. Place of residence: Cambridge, England. Living in a small house with her boyfriend and cat. Our paths were about to cross.

I had rented a cheap one-bedroom apartment just north of Cambridge city center, a perpetually damp and often sodden and chilling three-mile cycle to and from work. My windowless office was deep in the bowels of the University of Cambridge's Addenbrooke's Hospital. I was a research fellow in the Department of Psychiatry with no teaching or administrative duties. My job was to do pure research, and most of that took place in the newly established Wolfson Brain Imaging Centre, part of Addenbrooke's and a five-minute walk through a maze of corridors.

The Wolfson, as we all called it, was unique: its PET scanner was located right next door to the neurointensive care unit. Patients could be wheeled in their beds through two sets of swinging doors straight into the scanner. Indeed, the Wolfson's mantra in those early days was "Sick patients cannot go to the scanner, the scanner must come to the patient!" Neurointensive-care patients had usually suffered horrendous road accidents, massive strokes, or prolonged oxygen deprivation following a cardiac arrest or a so-called near-drowning incident. The easy proximity of the PET to the ward created a whole host of new opportunities for scanning bedridden patients with serious brain injuries.

It was a very different set of circumstances from Montreal and the Neuro, although both places had pros and cons. In Cambridge, the focus of my research was on brain injury. I wasn't *treating* patients like my colleagues, who were mostly MDs. Their day-to-day business was saving lives, administering treatments, and shepherding patients back to good health. In contrast, I was scanning them, trying to work out how their brain damage had affected their behavior and why. It was research of a very clinical kind. In Montreal, my research had been more about basic science, trying to understand how the healthy brain works and developing new techniques to investigate it. In an odd way, my experience at the Neuro had prepared me well for putting theory into practice in the intensely clinical environment of the Wolfson.

At the Neuro I had been able to touch a living human brain. It was normal practice for the resident neurosurgeons in Montreal to invite us mere scientists into the operating rooms to witness a person's life being held in their hands as they peeled back the skin, sawed away at the bone, and pulled back the meninges to reveal the trophy within—moving, pulsating, and alive, as vulnerable a sight as you are ever likely to see.

I ended up watching my first neurosurgical procedure up close in Montreal simply because I sat down next to one of the junior neurosurgeons in the canteen one day.

"You mean, you've never seen real brain surgery?" he said, perplexed that a young neuroscientist who spent his days peering at brain scans had never laid eyes on the real thing. "Come on down tomorrow, and I'll show you."

In crucial ways, my experiences in the operating room in Montreal taught me more than all my years of peering at brain scans. The most important lesson I learned is that your brain is who you are. It's every plan you've ever made, every person you've fallen in love with, and every regret you've ever had. Your brain is all there is. It's the pulsating essence of you as a person. Without a brain, our sense of "self" is reduced to nothing.

Without a heart we can live on with the help of machines. A patient with an artificial heart is still the same person. Without a liver or kidneys we can survive, personality unchanged, until the death of another soul provides us with a transplanted organ with which we can resume our lives, pretty much as we did before. We can lose arms, legs, eyes, and more and remain the same people, altered but nevertheless still *us*. Yet without our brains we are nothing more than a memory to others. We are not even a shadow of our former selves. We are gone. In the operating rooms of Montreal, I had learned the most important lesson in neuroscience—we are our brains.

I was never invited into the operating room in Cambridge, but something else was happening. In Montreal, the problems we tackled had been pure, basic science: "This is the equipment we have, this is what we know, let's put it all together and ask the next most important question about how the brain does its thing." We created the template, the hypotheses, and designed scans to fit. In Cambridge there was uncertainty. We were all over the place. We couldn't construct experiments beforehand. We had patients with types of damage to their brains that had never been scanned. There was no well-trodden path, no instruction manual or scientific map. There *was* opportunity. That was precisely the case with Kate.

≈

One June day in 1997 my colleague and friend Dr. David Menon—a gangly Indian neurointensivist with impeccable manners and infectious charm—told me about Kate. A bad cold had turned into a much more serious viral condition known as acute disseminated encephalomyelitis. Susceptible patients start to get neurological symptoms that may include confusion, drowsiness, and even coma. Kate was one of those patients.

The disease involves widespread inflammation of the brain and spinal-cord tissue and destroys what is known as white matter—not nearly as famous as gray matter but equally important. Gray matter refers to the outermost layer of the cerebral cortex. That's where all the action happens—where your memories are logged and your thoughts, plans, and actions germinate. Gray matter consists of countless neurons—specialized cells that relay nerve impulses.

White matter is the communication network between disparate gray-matter regions. White matter is mainly made of axons—dense tracts of highly insulated fibers, a kind of complex, supersophisticated cabling. White matter is white because of all its fat, or myelin, as it is more formally known. Fat is great electrical insulation. White matter enables areas of gray matter to communicate. These messages between neurons go much faster if the axons are insulated. Without insulation, the electrical signals literally leak out and the message is lost.

Kate's compromised white matter impacted her brain's communication network. She lapsed into a coma and was admitted to Addenbrooke's neurointensive care unit. Within a few weeks, she had improved. She had sleep-wake cycles, her eyes opened and closed, and she appeared to look fleetingly around the hospital room. But she showed no signs of inner life. No responses to prompting by her family or doctors. The infection was assumed to have left her completely unaware of who and where she was and what had happened to her. Doctors declared her vegetative.

I don't know why David and I thought of scanning Kate while she was in this vegetative state, but I can't help thinking that Maureen might have had something to do with it. It had been less than a year since her vegetative-state diagnosis and I was still very much coming to terms with her accident. Some part of me kept wondering what, if anything, might be going on in Maureen's brain. They said that she was vegetative, just like Kate, but what did being "vegetative" even mean? Perhaps Kate could help me find out.

David and I discussed what we might do with Kate. We came up with the idea of showing her pictures of her friends and family while she lay in the PET scanner. I knew a lot about which parts of the brain respond to familiar faces from my Montreal PET activation studies. We contacted Kate's wonderfully warm parents and asked them for ten photos of her family and friends. We told them we were going to try a new type of scan in an attempt to uncover what was going on in Kate's brain.

Kate's parents provided ten photos of people, all strangers to me. I ran them through a flatbed scanner, uploaded the images to my computer, cycled back to my damp flat, and spent the evening writing a simple program in Microsoft QuickBASIC that would present each image for ten seconds on a computer screen, one after the other. I also needed "control" images—photos that were as visually stimulating as the original photos but contained no discernible faces. I took each image, copied it, and defocused the copy using one of the early image editors of the day. Not a scientifically perfect experiment by any means, but fit for my purpose (fuzzy photos of human faces don't pass muster as an adequate control for real photos of faces). I was running out of time, and I didn't have the technical equipment to do anything more sophisticated.

David and I would show Kate the digitized images of her friends and family, and the unfocused versions of the same images, and look for different patterns of brain activity. If we saw a difference in the parts of Kate's brain that process information about faces, then I knew we would have discovered something important—that Kate, or at least her brain, could still perceive familiar faces.

To attempt to activate the brain of a vegetative-state patient was something completely new. Would her brain still respond to the faces of the people she had known and loved? Our question was that simple. We had forgotten, however, that before we could ask that question, we needed to determine if the visual information hitting her retinas was actually reaching her brain. What if the connection between her optic nerve and her cortex was severed or the information traveling along that pathway was interrupted? It would hardly be surprising if her brain failed to respond to the faces of people she had known. She couldn't see them!

We needed a quick solution. Kate might die or, less likely, recover. Either way, the scanning opportunity would be lost. I looked at the computer screen that we were going to use to show Kate the pictures of her friends, and in the delay, it had switched to screen-saver mode. It was 1997. Flying windows were all the rage. Red, blue, green, and yellow—they flew out at me, whizzing past, an intergalactic figment of a Microsoft engineer's imagination. We would show Kate the screen saver! The fast-moving, colorful display was perfect for checking that information was getting from her eyes to her brain.

As Kate lay in the scanner, we let the screen saver do its work: hitting her retina, firing up her optic tract, and activating her visual cortex. Then we let her rest—turned off the screen saver, placed a cloth over her face to shut out all light, and scanned her again. We did that several times. Screen saver, cloth, screen saver, cloth. At the end of the session we had what we were looking for. Kate's visual cortex sprang to life whenever we showed her the screen saver and returned to relative inactivity when a cloth covered her face. Visual information was reaching Kate's brain. Her brain, at least, "could see."

It was time to ask the big question. We flashed the two sets of images, faces and fuzzy faces, on a monitor suspended over the scanner bed. Kate was wheeled back to her ward, and we set about analyzing the data. We didn't know what to expect, but when we had the results in hand, we were stunned. Kate's fusiform gyrus had responded to the faces, crackling with activity. Moreover, the activity pattern was strikingly similar to what we, and others, had observed in people who were healthy and aware.

We felt like astronomers looking for extraterrestrial life who had sent a beep deep into outer space. Except in our case we were sending a beep deep into *inner* space. And a beep had come back! We'd made first contact. But what did it mean? Was Kate actually conscious despite her outward appearance? This question would perplex us for almost another decade.

There were no easy answers. Consciousness usually comes in two flavors, wakefulness and awareness. When you're put under with a general anesthetic, you plunge into what resembles sleep. That's you losing *wakefulness.* You also lose any sense of where you are, who you are, and your predicament. That's you losing *awareness.*

The wakefulness component of consciousness is relatively easy to understand and measure—if your eyes are open, you're awake. Awareness is much more difficult. How do you measure it? Gray-zone patients such as Kate illustrate this point perfectly. She was awake—there was no doubt about that—because her eyes were wide open. But was she aware?

Because Kate didn't respond to the sights and sounds around

her, or any of the numerous attempts to attract her attention, clinically the conclusion had been that she lacked consciousness. Her sense of self had been obliterated. A bit like an Alzheimer's patient, late in the course of the disease, who no longer has any sense of who or where she is. But Kate's predicament seemed even worse. Alzheimer's patients (at least until the very last stages of the disease, when they might enter a form of vegetative state) still retain a sense of *being something*, even after the sense of somewhere or someone is long gone. A connection exists with the outside world, although it is woefully weak and distorted. We had assumed Kate's connections were severed, utterly and entirely. That she had no sense of *being anything*.

Now we had new information. Our imperfect little experiment told us something vitally important. When Kate was shown pictures of people she knew, her brain responded just as if she was awake and aware, just as if she was a perfectly healthy person. What were we to make of this brain response? Could we equate it with the experience that she, as a person, might be having at the time? Did Kate experience the memories and emotions that we all typically experience when presented with a photo of someone we know and love? Did she *know* she was lying in a PET scanner, viewing photos of family and friends? Or was her brain responding automatically, as if on "autopilot" while she lay blissfully in "wakeful unawareness"?

Many types of stimuli—including faces, speech, and pain—produce automatic brain responses, echoes indicating that the message has been received though not necessarily consciously *experienced*. At a noisy party, we might be entirely unaware of a conversation going on over our right shoulder until the moment we hear our name. *This* grabs our attention. That we hear it at all must mean that despite having no conscious knowledge of our doing so, our brain has been monitoring that conversation just in case something important, such as our name, crops up. This doesn't mean that, because we perceive our names, our brains will remember the conversations in which they occurred. Memory and perception are entirely different. Perceiving a conversation doesn't mean that you'll remember it. Why would you? What's the point? What the brain is doing is scoping around, trolling for relevant information. It's not trying to remember everything.

The same thing happens with faces. As we walk through a crowded street, the familiar faces of our friends and acquaintances literally hijack our consciousness from whatever we were thinking about at the time. We notice, or as psychologists say, we *divert* our attention. That this happens tells us that our brains must be monitoring all the other faces, deciding which are worth attention and which can be happily ignored. But we're not conscious of doing this. It just happens. Our brain unconsciously sorts through the crowd, only alerting us to those people we might want to know are there—those that we recognize. Even if we try to control this process, we will fail; we cannot decide *not* to recognize a familiar face, no more than we can decide not to hear our own name at a party.

This phenomenon depends on where we are and what we're doing. On a street crowded with strangers, the faces of our friends grab our attention. But at a party full of friends, it's the stranger—the *unfamiliar* face—that we notice. This has to do with context and expectation and likely relates to the evolutionary advantage of being able to spot what is important from the barrage of information constantly hitting our retinas. On a crowded street, we don't expect to see people we know; it's a violation of expectancy and causes the brain to jump. This is fortunate. Running into friends among strangers is a good thing. It's adaptive. It might lead to a conversation, a date, a love affair, a partner for life.

Conversely, at a party full of familiar people, the stranger is the most interesting. We expect to see our friends there; an unfamiliar face violates that expectancy. We know all about our friends. But the stranger in the room? That could lead to something new. Again, it's adaptive. In every context, it's important to spot the different and unexpected. Our brains are highly efficient at spotting the odd one out, and most of the time they do this without our even knowing it.

Many of our brains' most sophisticated processes are like this. As adults we can't decide not to understand something that is being said to us. We can't decide not to learn how to get home from work if we travel that route every day, and we can't decide not to like a particular piece of music or art. We can decide not to *say* that we like it or even to declare that we hate it; but that doesn't change the underlying emotion, which is not our choice to experience.

In other words, many aspects of how we think and feel occur despite our having absolutely no awareness that these things are happening. By the same token, "normal" neural responses to events in people in the vegetative state do not necessarily mean that these people have any *conscious* experience associated with those events. This doesn't mean that they are *not* conscious either—conscious people also generate those same responses. All it means is that we just don't know. As revolutionary and exciting as Kate's response in the PET scanner had been, we just didn't know about her either.

None of this stopped us from thinking about it and talking about it. When our paper describing Kate's extraordinary case came out in the *Lancet*, one of the world's oldest (1823) and best-known medical journals, there was a flurry of media attention.

My colleague David Menon and I appeared on BBC morning television. I sat nervously in the studio, pointing at a life-size plastic model of the human brain and explaining the function of the fusiform gyrus. David added, "Imagine what would happen if an injury to the brain, or a disease that affected the brain, was [such that] not even eye movements were possible. If we didn't get a response from the patient, we wouldn't know if they were not responding or were not able to respond. It's truly a nightmare scenario."

Looking back at the grainy footage, I am struck by the strange set of coincidences and luck that had led us to that point. If Maureen hadn't had her accident, I might not have had any interest in the vegetative state; I might not even have known what it really meant. But wondering what might be going on in the brains of people such as Maureen had sown a seed of interest, and Kate had given me an opportunity to start experimenting. And then, what if Kate's brain hadn't responded? What if she had fallen asleep? Our response to this "look and see" experiment might well have been "Oh, well, that's not worth trying again. Let's move on and do something else." By some amazing stroke of luck, she was one of the few who *was* in there. It was she who gave us the impetus to look for others like her. I couldn't help but wonder whether Maureen might be in there too.

≈

Some months later, Kate began to recover and was moved to a specialized rehabilitation facility in one of the villages outside Cambridge. I was kept apprised of her progress. She gradually began to answer questions, read books, and watch television. Her thinking and reasoning skills were within the normal range, although she remained severely physically disabled. Parts of her brain that controlled walking and talking had been damaged.

Why did Kate recover? The medical thinking at that time was that patients diagnosed as vegetative for months on end *never* recovered. Did those people who cared for Kate change their behavior and attitudes toward her in light of our scan? Did they pay more attention, invest more time in her rehabilitation, and push her harder? Did this contribute to her recovery? Psychological studies have shown the devastating effects that social isolation can have on the brain. Imagine being ignored and treated like an object for days, weeks, and months on end. Surely that's the worst kind of social isolation. How could anyone come back from that? What a relief it must been for Kate to be talked to, read to, and included in every conversation. We don't know what effect that would have on the brain, but there's little doubt that it would have been empowering.

≈

Kate's recollections about her vegetative episode are harrowing. "They said I could not feel pain," she has written about her ordeal. "They were so wrong."

She was terrified when mucus was removed from her lungs. "I can't tell you how frightening it was, especially suction through the mouth." A raging thirst often gripped her that she couldn't signal. Sometimes she'd cry out. The nurses thought it was a reflex. They never explained what they were doing to her.

Kate tried to take her own life by holding her breath, an all-too-common strategy for conscious people in the gray zone. "I could not stop my nose from breathing. My body did not seem to want to die."

Making first contact with Kate and her subsequent recovery generated more questions than it answered. When did she become aware? What parts of the brain are essential in that process? Which are ancillary?

I felt as though we had ventured into the underworld and convinced someone there to follow us back out. It seemed Kate felt that way too. She wrote to me some years after we'd first scanned her, when she was back living with her parents in Cambridge:

> Dear Adrian,
>
> Please use my case to show people how important the scans are. I want more people to know about them. I am a big fan of them now. I was unresponsive and looked hopeless, but the scan showed people I was in there.
>
> It was like magic, it found me.
>
> Love from Kate

Over the years, Kate and I stayed in touch, mostly by e-mail. Sometimes she'd write four or five times a week, and then there would be months without contact. I felt an enduring, close connection with Kate, something that had a profound influence on me and my work; she was always Patient #1, always the person I'd refer to when I gave lectures about how this journey began. We had each changed each other's life.

As I look back over those e-mails now, it's clear that despite her miraculous "recovery" Kate's life was far from easy. "Had a tough year, not nice at all. Had both big toes amputated and a really awful stay in hospital," she once wrote. It shocked me to read that. Then: "Sorry I was so down in my last e-mail, I had a very bad Christmas time so was feeling low."

The e-mails reveal her shifting moods. Yet between bouts of despair a gritty determination emerged. Kate endured despite all she'd been through. "I think my determination was the main thing that helped me. I always have been determined."

Then, in June 2016, almost twenty years to the day after her brain injury, I visited Kate in Cambridge. It was raining hard when I got off the train from Heathrow Airport. It always seemed to rain hard in Cambridge. And it was a chilly rain, the plague of British summers, which reminded me of growing up and rainy family holidays spent on the beaches of southern England. My baggage had been delayed in Toronto, and all I had was my old Canon camera and the clothes that I'd flown in, which didn't include a coat.

As the taxi wound through the narrow country lanes, I was apprehensive. It had been more than seven years since I'd last seen Kate, a year or so before I left the UK to return to Canada more permanently. She'd been living with her parents, Gill and Bill, and we'd caught up over tea as I asked her questions about her life and she responded, slowly and methodically, by pointing to letters on a board. As remarkable as her recovery had been, her speech was still quite impaired, and I couldn't make much sense of anything she said. I wasn't looking forward to going through this process again, communicating letter by letter, sentence by sentence, and I was quite sure that she wasn't either. But she'd agreed to meet me, and for that I was grateful and willing to do whatever it took to make it easy for her. Trying harder to understand her broken speech would be a good start, I thought.

My mood lifted as the taxi turned into Kate's street in a quiet, pleasant neighborhood on the outskirts of Cambridge, and it suddenly stopped raining. The sun burst through the clouds. A good sign? I noticed that Kate's house, like all the houses around it, was single-story. Wheelchairs and stairs don't mix. The house was what is called in the UK a council estate. Government-owned housing. Because Kate has no income and is on disability welfare, she doesn't pay rent and her living expenses are covered.

I rang the bell, and a cheerful care assistant opened the door, introduced herself as Maria, warmly shook my hand, and ushered me in. The National Health Service covers Kate's round-the-clock care.

Maria led me into the comfortable living room. There was Kate, ensconced in her electric wheelchair.

"Hello again!" I took hold of both her hands. "I bought you flowers!" I gestured toward the bouquet of lilies I had picked up.

"Thank you very much," Kate replied without missing a beat. "They're quite nice."

They're quite nice. I was stunned. Kate had just spoken. No letter board, no broken speech. *Kate could speak!*

"Your speech is amazing!" I blurted out.

"I taught myself to speak again!" She broke into a winning smile that gave away exactly just how pleased she was with herself. "I love to talk."

"Do you mind if I record our conversation?"

She gave me a glum look. "I hate hearing my voice."

After some playful back and forth, she capitulated.

"How did it feel when you first woke up after your period of unconsciousness?" I asked.

"I thought I was in prison. I had no idea where I was."

"What was the last thing you remember?"

"I was at school, where I worked as a teacher, having lunch. When I woke up, I didn't feel like I'd been asleep. I was just suddenly *there*."

"I thought you became gradually conscious."

"It was like that—just a short time in the beginning with a little bit more every day. Consciousness came back slowly. The very first time I was conscious all day I had an OT [occupational therapist] with me. She was called Jackie. She was the *only* person in those early days who told me her name and job. Very few people told me their names."

"Why do you think that was?"

"They thought I wasn't me; they thought I was just a body. It was horrendous. I still had feelings. I was still a person! I was incredibly angry inside. The main thing is I had no idea where I was or why I was there. I thought I'd forgotten how to walk."

"No one told you where you were?"

"I couldn't hear anyway. I could only hear noise. No words."

≈

Kate's story horrified me. I thought back to the time we'd scanned her, to the time we'd made first contact. With the benefit of hindsight it was now obvious that we'd stumbled upon something incredibly important all those years ago. Part of Kate was still there, and perhaps that's what was reflected in our early scans. In the weeks and months that followed, she'd been subjected to so many awful experiences, it was hard not to think that we might have done more to prevent that. Should we have tried harder to make sure that *everyone* treated her as a person? Should we have been more aggressive and issued directives to the staff and carers of all patients like Kate? We didn't know what we know now, and "sounding the alarm" in this way would have been premature; the result would have unrealistically raised the hopes and expectations of many thousands of families like Kate's. All we had at the time was the slightest hint that some part of Kate's brain was still working as it had done before her brain injury. Whether that meant she was aware we did not know, and to assume so would have been both unjustified and unscientific. Nevertheless, twenty years on, the thought that we could have done something more to alleviate Kate's suffering troubled me greatly.

Kate talked about the disease that had thrown her into the gray zone. "I'd love to know why I got it. I'm told I'll never know. Sometimes I think it must be my fault. God was punishing me."

"Are you a religious person?"

"No, but I have faith. I have faith in my head. I don't go to church. I didn't go to church before. I have never been religious. But I've found that faith has helped me a lot. It's hard to keep going. I need a reason. My brain won't give up. I can't cry. I've lost my tears, the ability to cry. It's horrendous. Really awful. One of the worst things."

I asked her what she meant by something she had said to me in one of her first e-mails: that the scan had "found" her.

"The scan found me inside. I was unconscious. I think I really wanted to sleep because my brain had to work extra hard to see." I thought, perhaps, Kate was referring to being directed by me to look at the photos in the scanner, and my impulse was to ask her about that, but I didn't want to interrupt her train of thought. "Even now I find it's really hard to watch films. I can watch the first hour or half an hour, and then I fall asleep. I can't wait for the new *Bridget Jones* film. I *love* my Kindle. I've read loads of books. I don't read modern books. I read old books. I love Jane Austen. Her heroes are lovely. Modern books remind me of what I've lost. My brain keeps going. My recovery is because of my brain. I thought I would just give up, but my brain won't give up. I fight my brain every day. It won't do what I want. It won't do what I ask."

"What do you mean by that?"

"My brain makes my body do things that I don't want to do. Like when my leg spasms. It doesn't like me. My brain doesn't like me. It won't give up. It got cross with me. Before this I felt like one person, now I feel like two. The old me, before I got ill, was a different person. I feel like I died. And now I'm alive again."

Kate spent quite a bit of time talking to me about this strange sense of duality: her feeling that the person she was now was not the person she used to be. In one sense she was quite right: many aspects of her life had changed beyond recognition; but for the most part these were physical changes. I wanted her to tell me that her mind, the part of her that defined who she was, was unchanged. That she had returned from the gray zone bruised, perhaps, but mostly intact. But for Kate, it seemed quite the opposite. Even her own brain, she felt, was working against her. Something about Kate had changed, something about her had been lost in the gray zone.

I asked Kate if there was anything she wanted to say, something I hadn't asked.

"The most important thing to remember is that I'm a person, just as you're a person, and I have feelings, just as you have feelings."

I left Kate and took off down the driveway to my waiting taxi. As we pulled out of her quiet suburban street toward the hustle and bustle of Cambridge, it began to pour again. I couldn't help thinking about everything I'd learned from Kate. The gray zone is a dark place, but she'd shown me that it *is* possible to come back. The human brain has amazing power to heal itself. Kate also taught me that the essence of a person, the "me" in me, can survive the worst of times. Her spirit was unbroken despite her travails.

CHAPTER THREE

THE UNIT

≈

> King Arthur: "Cut down a tree with a herring? It can't be done."
>
> —*Monty Python and the Holy Grail*

Soon after I arrived in Cambridge, back from Montreal, I formed a band called You Jump First with a group of academic friends and started to play gigs in pubs around Cambridge. I sang and played bass at the same time, which was a terrible idea. Few people have pulled that off (Sting, Paul McCartney, and a few others). I quickly switched to acoustic guitar, and we found our sound—Celtic-infused pop-rock with a splash of Bruce Springsteen. We entered band competitions, both locally and farther afield. One of these competitions took place in Hertford, a small town in the south of England, not far from St. Albans, where Maureen's brother Phil lived. He was a computer scientist working on software development for 3Com. Tall and slim, he reminded me of Maureen: they had exactly the same teeth. I invited him to come along, and he turned up to cheer us on. After we came offstage, I asked him about Maureen.

She was still living a few miles from her hometown of Dalkeith, which is near Edinburgh in Scotland. Her parents hoped to move her to a more local nursing home in the coming months. Otherwise, Phil said, there was nothing new to report. Almost two years had passed since Maureen's injury, and I was beginning to wonder if she would ever recover. I told him about Kate and how excited I was about the results of her scan and the possibilities it held for patients such as Maureen. We promised to keep in touch.

≈

Writing about Kate's case in the *Lancet* in 1998 had been a milestone for Cambridge and a significant change in scientific direction for me. I was entirely unsure where it would lead. I had no funding, beyond my own salary, and no lab as such; it was just me in an office with a computer. I was utterly dependent on the goodwill and research grants of those around me.

Then serendipity dealt me a game-changing hand. I was offered a job at the Applied Psychology Unit of the Medical Research Council (MRC), a government agency that funds medical research in the United Kingdom—medical research that has produced thirty Nobel Prize winners to date. My position at Addenbrooke's Hospital was for three years, after which the funding for my salary was destined to run dry. The gig with the Unit was open-ended, and the prospect of a permanent job and eventual tenure was too much to resist.

The Unit was established at Cambridge in 1944 and for over half a century had a very British influence on psychology. Its day-to-day business of making scientific breakthroughs in our understanding of memory, attention, emotion, and language was interrupted twice a day for tea in the common room and, weather permitting, croquet on the lawn. Indeed, the Unit had on its payroll an elderly gentleman called Brian, stooped, with thinning white hair, whose principal job was to make tea and coffee, which was ceremoniously served up on an equally ancient cart universally known as "the tea trolley." On special occasions, such as somebody's birthday, we'd also get cookies, but for the most part it was just tea and coffee, once in the morning and once in the middle of the afternoon. It was never clear what Brian did in between the morning and afternoon tea-trolley runs, and I never thought to ask. The common room, where tea was served, still resembled the grand old parlor that it presumably once was, with a large fireplace long since retired, ornate ceiling moldings, and a lonely looking centerpiece that had probably parted company with its chandelier half a century before. The Unit's Christmas pantomime was legendary—a very British tradition in which men seize every opportunity to slip into a dress, lipstick, and a wig and become their favorite female bombshell. Having spent my formative years at Gravesend Grammar School for Boys, where such events were commonplace, nothing about the Unit upon my arrival seemed in the least bit strange to me.

The Unit was headquartered in a huge Edwardian manor on Chaucer Road, a quiet leafy street just south of Cambridge city center. It had begun life as part of Cambridge's Psychology Department. But by 1952 the third director, Norman Mackworth, found it had outgrown the space available within the department. He noticed a pleasant old Edwardian manor house on the outskirts of the city with a large garden and lawns ideally suited to croquet, bought it with his own money, and informed the MRC that this was to be the Unit's new premises. I'm sure such things only happen in Cambridge.

By the mid-1960s the Unit was staffed by clean-cut scientists, almost all male, of course, strutting about in tweed jackets and ascots, smoking pipes, twiddling knobs and occasionally sipping a glass of sherry. It was a very British way of doing science, and Cambridge in the 1960s was about as British as the British could get. It's hardly surprising that by decade's end Cambridge had produced half of the Monty Python team.

Work at the Unit often resembled a Monty Python sketch. One test I administered was designed to measure "perseveration"—a problem with attention that makes you keep doing the same thing over and over, even when you are told not to. My patient had sustained frontal-lobe damage. I asked him to name as many words as he could think of beginning with the letters *F*, then *A*, then *S*. Most people without brain damage produce such words as "face, field, fox, falcon, frost . . ." until they run dry. My patient started with "Five, fifteen, fifty, five hundred." I realized I was in for a long day as he continued, "Five hundred and one, five hundred and two, five hundred and three—"

"Stop!" I said. "Let's try something else. Have a go at *S*."

Quick as a flash, he exclaimed, "Easy! Six, sixteen, sixty-six . . ."

≈

By 1997 a strong division—almost a tension—existed between the Unit and the Department of Experimental Psychology, where I had previously worked as a research assistant between 1988 and 1989. Both were eminent Cambridge institutions, but the focus of each was quite different. At the Unit, you might study how it was that we remember sequences of digits. Most of us can listen to and correctly repeat "digit spans" of a five- or six-number sequence. We are able to increase that number with techniques such as chunking—splitting, say, the number 362785 and remembering 362 followed by 785.

Similarly, it's easier to remember a longer sequence if it involves repetition. We easily repeat twelve digits if they take the following form: 497497497497—all we have to do is remember that the sequence 497 repeats four times. Our brains are great at spotting repetition, or chunking information into memorable packages, often without our being entirely aware how it's happening. We *are* aware that it is happening, but it mostly occurs automatically without our even knowing it, an unconscious process that we can become of aware of, but often only after it has occurred.

Through a series of clever studies at the Unit, my former student Daniel Bor showed that this memory recoding, where information is repackaged and organized to make later retrieval easier, is carried out by regions of the brain that have been linked to general intelligence, otherwise known as g, which is measured by tests of IQ. This all makes a lot of sense if you think about it. Being "intelligent" depends on much more than memorization. It depends on what we do with what we remember, how to make what we remember *useful* in a variety of ways. And that has to do with the way we lay down memories, the way we organize and catalog them, and how easy it then is for us to retrieve them efficiently. How we organize our memories impacts almost every aspect of cognitive function and gives some of us a competitive advantage in almost every aspect of life that depends on it. Chunking numbers and letters is the most simplistic form of this process. But learn to do it and you'll be better at remembering phone numbers, license-plate numbers, addresses, and much, much more. As Ella Fitzgerald once sang, "'Tain't what you do; it's the way that you do it."

Both the Unit and the Department of Experimental Psychology studied how we organize memories, but their approaches diverged. In the department, you'd be more likely to study working memory and phenomena such as chunking from a different angle, looking at why the loss of dopamine in the basal ganglia of patients with Parkinson's disease compromised working memory, or studying how drugs such as Ritalin can improve working memory in healthy people.

These two worlds, which we might call the psychological versus the neuroscientific, were colliding and combining in 1997 when I arrived at the Unit. Cognitive neuroscience—which combined aspects of psychology, neuroscience, physiology, computer science, and philosophy—was the new hot field. It provided a legitimate platform from which professionals not medically trained (non-MDs such as myself) could study many different types of patients in the pursuit of scientific knowledge.

I was hired by the Unit to spearhead its forays into brain imaging through my established connections with the Wolfson. The Unit didn't have a scanner; that was based back at Addenbrooke's Hospital in the Wolfson Brain Imaging Centre. But the Unit was flush with the new breed of cognitive neuroscientists desperate to get their hands on a scanner and start asking probing questions about the human brain. A deal was struck; the Unit would pay the Wolfson for time on their scanner, and I would be responsible for booking scans, allocating time, deciding who got access and who didn't, and basically keeping the whole system ticking along nicely. So in July of 1997 I relocated to the manor on Chaucer Road and gained immediate access to its pot of research funds. Every five years, up to 25 million pounds was at stake: all of our salaries, expenses, not to mention the manor's heating, lighting, Brian and his trolley, and the small army of gardeners who kept the croquet lawn groomed.

I was quickly surrounded by people who harbored a similar passion for understanding how the brain works and, perhaps more important, for using every fancy new tool they could get their hands on to push the bounds of neuroscience. We were intoxicated by the power these new brain-imaging tools gave us. We thought we would soon be able to tell the world what made each of us who we are, what made us *us*! All this plus tea, crumpets, and sticky wickets. The Unit, with its eccentricities and dry, understated British humor, was a perfect scientific milieu for me to work out where we would go after Kate.

And then along came Debbie.

CHAPTER FOUR

HALF-LIFE

≈

Your every thought is a ghost, dancing.

—Alan Moore

Debbie was a thirty-year-old bank manager who had been trapped in her car after a head-on collision, her brain starved for oxygen, a dire situation that occurs surprisingly often. In intensive care at Addenbrooke's she had unreactive pupils, a bad sign that indicates injury or compression of the third cranial nerve and the upper part of the brain stem.

Even slight brain-stem damage can be catastrophic, disrupting sleep-wake cycles, heart rate, breathing, and consciousness itself. Sensory signals related to hearing, taste, and the sense of touch and pain will be disrupted to the thalamus, a central relay station or hub. A small amount of damage to the brain stem can put you in a coma. Many neurosurgical patients I saw as a PhD student had huge chunks of cortex—sometimes the size of a tangerine—surgically removed to alleviate epilepsy or excavate a tumor. Afterward their mental faculties were only subtly affected. Enormous parts of the brain can be damaged or removed entirely and cause minimal disruption, while a tiny lesion in a critical hub such as the brain stem or thalamus can be devastating.

Fourteen weeks after Debbie's accident her pupils were still dilated and unreactive. She was doubly incontinent, fed through a plastic tube inserted into the stomach, required twenty-four-hour nursing care, was completely nonresponsive, and was declared to be in a vegetative state. Her family felt, however, that when well rested she would occasionally respond to them. We found no evidence of response at the bedside. She flinched to painful stimuli, such as pressure applied to a fingernail. But such responses are reflexive, common in patients in the gray zone, and don't necessarily signal awareness.

Quickly removing the hand that you accidentally placed on a hot stove is automatic and instantaneous and involves only the neurons in your spinal cord, not your brain. It would simply take too long if the message "Hot!" had to go up your arm to your spinal cord and then on to your brain for you to *decide* to move your hand, only to then send that message back down to your arm. Painful stimuli such as pressure on a fingernail or the feeling of a hot stove elicit a hardwired, automatic response, which tells us little about patients in the gray zone: these responses occur whether or not the brain is irreparably damaged.

We scanned Debbie in 2000 twelve times. Each scan lasted ninety seconds, the optimal duration for acquiring the best images of the functioning brain before the radioactive tracer O-15 (known as oxygen fifteen) decays to a level that is too low to detect.

Like most radioactive materials used for medical treatment and research, O-15 is produced by a cyclotron, a type of particle accelerator, which was buried in Addenbrooke's basement behind thick concrete walls to keep radiation in and people out. The radioisotope was pumped upstairs to the Imaging Centre and administered to Debbie through an intravenous line inserted into her arm as she lay in the scanner.

O-15 has a half-life of 122.24 seconds, not much longer than the length of a single PET scan. But with this method, each scan provides an image of blood flow, averaged over a period of ninety seconds from when the tracer first enters the brain. Once it enters the bloodstream, the O-15 is pumped to the right side of the heart, then to the lungs, back to the left side of the heart, and finally to the brain, a process that takes fifteen to thirty seconds—a constantly decaying river of radioactivity poised to reveal the brain's secrets.

We were applying the same technology we'd used in Montreal. During scans, some parts of the brain work harder than others, depending on the scanned patient's thoughts, actions, or emotions. Brain areas working hardest quickly deplete energy in the form of glucose, which has to be replaced so that those brain regions can go on exerting themselves. The brain sends more glucose to those areas via blood. Active areas attract more blood, and because the blood has been labeled or tagged with radioactivity, the PET scanner sees where it goes.

Our primary question, which we pondered for several weeks, was what to do with Debbie while she was being scanned? How should we try to activate her brain? I thought back to the day that David Menon and I had scanned Kate and remembered that during three of her twelve scans we could see through the window of the control room next door that her eyes were closed and she appeared to have fallen asleep. She couldn't have seen the photos of family and friends. The remaining nine scans had luckily produced convincing evidence of a brain response. But what if Kate had fallen asleep for most of or even the whole session? What if she had deliberately or inadvertently closed her eyes? We had waited for three years for another chance at scanning a patient like Kate. Three years of wondering whether she was a one-off. It was exciting and harrowing—we couldn't mess up.

You may wonder why it took three years to finally scan another vegetative-state patient. First off, we were slowly developing the methods that we would go on to use to probe the gray zone. What was the right thing to ask the person to do in the scanner, and should it be the same for everyone? With no funding to support this kind of work, I was spending most of my time working on other projects: how the frontal lobes function and why patients with Parkinson's disease have cognitive deficits. Also no "system" was yet in place for getting patients to us from other hospitals, so suitable candidates had to end up at Addenbrooke's for us to know they existed. And even if I had known about patients at other hospitals, who would have paid to transport them to me?

As we tried to figure out the kind of experiment we should do with Debbie, we knew we needed to move quickly. She might die, lapse again into a coma, or become hooked up to machines that would make scanning her impossible. Trying to activate Debbie's brain through her visual system, as we had done with Kate, felt risky. Then it occurred to us to use *sound*. You can close your eyes but you can't close your ears! During six of the ninety-second scans, we would play Debbie a series of words through headphones.

These were not ordinary words. At the Unit, I found myself surrounded by psycholinguists, language experts, who knew what words we would need to produce brain activity that we could interpret with confidence. Carefully controlled words that were not too abstract, yet abstract enough to elicit a mental representation; not too familiar, but familiar enough that memories would be evoked related to the content of those words.

My new psycholinguist friends knew about the relationship between language and the brain, which parts of the brain processed which aspects of language, and which types of speech stimuli would produce certain patterns of brain activity. If someone speaks to you in a foreign language that you've never before heard, what does it sound like? Noise? A lawn mower? Of course not! It sounds like speech spoken in an unintelligible language. But how does your brain know that it's speech and not just noise?

The answer is that the brain has specialized modules in the temporal lobe that are good at determining what is and is not speech, even if it is presented in an unfamiliar language. That is why it's impossible for us to tell the difference between the made-up languages in TV shows such as *Game of Thrones* and real languages with which we have no prior experience. Both sound like language, both are equally unintelligible, and our brains classify them in the same way. But neither sounds like a lawn mower, and our brains know that because of a specialized "speech detection module" that sits high on the temporal lobes, large cortical regions situated on each side and toward the bottom of our brains. The upper part of these lobes is devoted to processing sound, which is why it's often referred to as auditory cortex. And a specialized region within auditory cortex, the planum temporale, is specifically devoted to processing speech sounds. It detects speech and tells the rest of the brain that speech is what it is hearing.

The words we played to Debbie were recorded on cassette tape. They were all two-syllable nouns (such as *sofa*) that had been carefully matched for how often they occurred in regular speech, their degree of abstraction, and how easily the described objects could be imagined. For example, it's easy to imagine a sofa, but a whole lot harder to visualize *uncertainty*, although both are common nouns. Absolutely everything—every word, when it happened, how loud it was, how frequently it occurred in the English language—was scrupulously selected. All I wanted to know was whether Debbie's brain lit up when she heard speech. Did the words we played to her all need to be two syllables and occur with precisely the same frequency in the English language?

I was told that all of these things were essential factors to "control" in our experiment. Even the rate at which the words were presented had to be measured by a metronome. My new friends were control freaks, and my experiment began to feel like one more Monty Python sketch. Fortunately, after several years in the Unit, I was used to it. And it was not just speech that had to be controlled. For six of the twelve scans Debbie heard short intervals of noise. Again, these were not any old noises but exquisitely controlled, carefully produced "bursts," called signal-correlated noise, which sounded like the static from an old radio that leaps out at you from between the stations as you turn the dial. Signal-correlated noise is similar except that, just like speech, it varies in terms of its amplitude (volume) and spectral profile, which is the combination of frequencies being played at any one time. It almost sounds as if that radio static were talking to you, except it's impossible to make any sense of what is being said.

Finally, we were ready. Debbie was positioned in the scanner, the intravenous needle inserted into her arm, and the O-15 flowed. The technician started the virtually silent scanner. Debbie didn't move. Nothing changed. Just the slow, insistent voice in the scanner room: "Sofa . . . candle . . . table . . . lemon." Two seconds between each word, then carefully calibrated bursts of noise. Debbie was wheeled back to the neurointensive care unit, and we set about trying to make sense of the data.

≈

In those days, it could take up to a week to analyze PET scans. Waiting patiently for outcomes left plenty of time for speculation and croquet. We encamped on the manor's lawn in Chaucer Road, sipping tea, and wondered whether we had been able to bring Debbie's brain back to life, and if we had, what would that mean? The week it took to get the results of the scan felt like a year.

When the results did finally appear on my computer screen, I was stunned. Despite Debbie's vegetative-state diagnosis, her brain responded to speech and noise bursts just like yours or mine. It almost felt too good to be true. First Kate; now Debbie. Both of their brains responded as if they were normal, healthy volunteers in one of our studies. Yet both were, apparently, in a vegetative state. Could it be that they were not vegetative at all but in there and fighting to get out? And if so, what did that mean for the other people all over the world in their condition?

Although we could not be certain that Debbie was conscious, we had shown that human speech could activate the vegetative brain. It was a thrilling result, and the Unit buzzed with excitement as we pondered the repercussions. My close friend and colleague John Duncan was amazed.

"I thought it would never work!" he said.

"Who knows?" I replied. "Perhaps she understands everything going on around her."

William Marslen-Wilson, the Unit's director, was less optimistic: "It could just be an automatic response."

He was right, but it nevertheless gave us plenty to think about as the annual summer croquet tournament reached a fever pitch. What we did know for sure was that we were starting to reveal the secrets of minds that no neurologist, however experienced or smart, could ever know through standard clinical investigation. It felt as though we were right at the beginning of a completely new interface between science and medicine.

≈

When we wrote about Debbie's case in the scientific journal *Neurocase* late that year, we sat firmly on the fence. We had to—there was still so much we didn't know.

One possibility, we pointed out, was that Debbie was not actually in a vegetative state at the time of her scan but was recovering—not to the point that it would be noticed at the bedside, but enough for her to activate her brain in our PET scan. Perhaps Debbie had been at least partially aware despite her vegetative-state diagnosis. A second possibility we discussed was that Debbie was another vegetative patient who showed limited fragments of brain function in the absence of any apparent evidence of awareness.

In part, we were responding to the results of a scientific paper that had been published in the *Journal of Cognitive Neuroscience* a year or so after our paper about Kate appeared in the *Lancet*. Its author was Dr. Nicholas Schiff of the highly esteemed Weill Cornell Medical College on Manhattan's Upper East Side.

In 1998, a few weeks before our paper in the *Lancet* was published, Dr. Schiff accompanied his mentor Fred Plum to Cambridge. Plum was a giant in the field of brain injury, and when we met, it was clear that Plum and Schiff's interests and ours were closely aligned. They told us about cases of their own, similar in some ways to Kate, but, then again, completely different. An odd paradox in gray-zone science is that patients are lumped into categories such as vegetative state that give the illusion that they are somehow quite similar while in reality every patient is completely different.

Schiff and Plum told us about a forty-nine-year-old American woman who had been unconscious for twenty years following three hemorrhages from a deep arteriovenous malformation of her brain. From time to time their patient (unlike Kate) would show fragments of behavior—infrequent isolated words unrelated to anything going on around her. A PET scan revealed islands of modestly higher metabolism than would be expected in someone who was unconscious, particularly in the areas of the brain known to be involved in speech. They concluded, "The presence of isolated modules of processing in patients diagnosed as vegetative state cannot be taken by themselves to enable any degree of self-awareness."

They were treading cautiously, sitting firmly on the fence—just like us. It was early days. What else could we do? But the title of their paper, "Words without Mind," disclosed a rather less optimistic view of these early imaging findings than we had on our side of the pond. Perhaps it was not just the results of Kate's scan but the ensuing publicity about her case and her startling subsequent recovery that had filled all of us with hope and wonder. And Debbie had only added to that thrilling sense of expectation and possibility.

≈

Schiff and Plum and their colleagues at Weill Cornell were not the only ones running head to head with our quirky little group in Cambridge. Important research in gray-zone science was emerging from Liège, a small university town in Belgium. A young neurologist named Steven Laureys was also beginning to explore the possibilities of using PET to investigate brain function in the vegetative state. In one of their early papers, Laureys and his team described the scans of four vegetative-state patients. Their brains appeared to be less tightly "connected" than healthy controls, with disorganized or fragmented patterns of overall activity.

Yet more evidence. A different kind of evidence, but evidence all the same. In Cambridge we were seeing vegetative patients responding normally in the scanner, despite showing no outward signs of consciousness. In New York and Belgium, fragments of behavior and patterns of brain activity in the vegetative state.

Gray-zone science as a field was beginning to coalesce. And the same year that we published our paper about Debbie, Dr. Joe Giacino and colleagues published a landmark paper describing, for the first time, the minimally conscious state. According to that report, many patients who appear to be vegetative might, in fact, be in a minimal state of consciousness, partly there and partly gone, occasionally able to signal their diminished awareness but never able to marshal these fragments of consciousness to communicate effectively with the outside world.

When you're half-awake, half-asleep, and someone says to you, "Please squeeze my hand," perhaps you'd do it or perhaps you wouldn't. You might hear the instruction but fade away before responding. Or perhaps you'd respond, although the next time someone said, "Please squeeze my hand," you'd miss it completely because by then you'd be fast asleep.

We don't know that is what it feels like to be in a minimally conscious state, but clinically that is how patients behave. Sometimes there, sometimes gone. It's a bizarre place that is different from that inhabited by vegetative-state patients. A less consistent, murkier place with patches of light and dark. With Giacino's new paper, we now had a whole new diagnostic category. A patient need be neither conscious and aware nor vegetative, but trapped in between in the minimally conscious state.

≈

We needed another scan to assess Debbie, but unfortunately she had reached her radiation burden. Unless we could make a strong case that another PET scan would benefit Debbie directly, our local ethics committee—who ultimately decided for every scientific study what we could and could not do—would not allow us to give her more radiation. And we couldn't make that case.

Although we knew we were onto something important, we could hardly argue these experiments would directly benefit Debbie. This was scientific exploration in its infancy. We were a million miles from clinical benefit.

Astonishingly, like Kate, some months after her scan Debbie began to recover. Quite quickly, she was given the new diagnosis of minimally conscious state that Joe Giacino and his colleagues had introduced. But again, like Kate, Debbie went beyond that; when I saw her a year or so after her scan, she had severe disabilities but was rapidly improving, starting to speak again, to move her limbs, and to return from the gray zone. She would pull herself up in her chair and laugh at her favorite TV programs, look at us when we spoke to her, and respond with fitful bursts of garbled speech that gradually became more and more intelligible. I lost touch with her when she was moved away to a long-term rehabilitation facility near her family home, and it became impossible for me to track her progress.

I often wonder about Debbie. Did we find a way to bring her back into our world? Did our scan and the flurry of attention it generated locally somehow contribute to her recovery? With both Kate and now Debbie, did our scans make people treat them differently and somehow, in other ways we were not aware of, help them get better? We didn't have enough evidence to be certain of anything. But their remarkable recoveries were beginning to feel more than coincidental.

CHAPTER FIVE

SCAFFOLDS OF CONSCIOUSNESS

≈

> The gates of hell are open night and day;
> Smooth is the descent, and easy is the way:
> But to return, and view the cheerful skies,
> In this the task and mighty labor lies.
>
> —Virgil

As 2002 rolled over into 2003, several things were starting to trouble me. First, there was Debbie and her brain activity. It was frustrating that we didn't know what it meant. We'd played her a list of words and her brain had responded just as yours or mine would. It detected speech, not confusing it with other noises. I desperately wanted to know if her brain understood what those words meant. A damaged, unconscious brain might register the sound of speech and not be able to do much with that information. But could an unconscious person still understand the spoken word? In that context, what could "understanding" possibly mean?

It's a complicated question—at what level of brain function are you conscious? That question would be at the forefront of my voyage into the gray zone as interest in the field exploded over the next few years. Part of the problem is that questions about consciousness have as much to do with personal taste as science.

Take the example of a young child. Most of us would agree that healthy ten-year-old children are conscious of themselves and the world around them in much the same way that adults are. They understand language, make decisions, respond to questions, lay down memories, act on stored memories, and have most of the other cognitive faculties of an adult, albeit in a more basic form.

What about two-year-olds? Are they conscious? Most of us would say yes. They understand language and make decisions, not complex ones, but whether to go play with a toy train or look at a picture book is a decision. They say words and sometimes entire sentences, store memories, and will, on occasion, act on those memories (retrieving a put-away toy train is acting on a previously stored memory). They have many of the basics of adult consciousness.

Now consider a one-month-old. Of course a one-month-old is conscious, you say! But think about it. One-month-old infants don't seem to understand what is said to them, although it might be possible to attract their attention momentarily with an "Ooh" or "Aah." If you scream at them (you shouldn't), they might start to cry; if you sing softly, they grow calm and perhaps coo. But that's about it.

Most of these "responses" are undoubtedly automatic, hardwired into the system from birth or even before. They aren't elaborate; in fact, they are rather rigid—singing softly will calm an infant regardless of what you are singing about. Infants don't respond to instructions with appropriate actions, but then they don't yet understand language, so let's give them a break there. They may or may not be laying down memories (few of us claim to remember being one month old), and they clearly don't appear to act based on remembered information in the way that a two-year-old will. They might orient toward a new toy, but once that toy is out of sight, it is gone from their world. So, are one-month-olds conscious? Do they "know" that they exist as a person and that there is a world out there that they can interact with, influence, and be influenced by? If they do, what form does that "knowing" take?

In short, it's a lot more difficult to decide whether one-month-olds are or are not conscious, and unsurprisingly, we are divided: some of us think they are; others aren't so sure. I debated this issue with the Dalai Lama in Brazil in 2010, and he gave the same answer that I get when I discuss it with my neuroscience colleagues: "It depends what you mean by consciousness." That's the problem! What mental faculties demonstrate consciousness? Debbie could detect speech, but that was insufficient evidence to conclude that she was conscious—at least to me.

Not everyone agrees with this logic. Ask your friends and you will quickly find someone who is completely sure that a one-month-old child is conscious (perhaps you are too?). But then try them with this: What about a fetus? Is it conscious? Even your die-hard consciousness friends might begin to have doubts. Let's push back further. What about a zygote—the single cell formed from the sperm and the egg that leads, nine months later, to the birth of a child? Is a zygote conscious? Most people will agree it isn't, in part because it has none of the capacities of an infant; it's also implausible that a single-celled organism could be conscious.

This raises an interesting problem. When, then, on this developmental trajectory from zygote to fetus to newborn to toddler to adult, does consciousness emerge? It doesn't matter whether you think a one-month-old (or even a fetus) is conscious. If you agree that a single-celled zygote is unlikely to be conscious, but a healthy adult is, then somewhere between these two points we must become conscious. But when? Birth is an obvious and dramatic change point, but it seems highly unlikely that a child fresh from the womb is any more conscious than a nine-month-old fetus about to be born.

We have no agreed point at which a developing organism, in this case a person, can be said to become conscious. It's easy to decide that a ten-year-old is conscious and a zygote isn't. But in between? A one-month-old exhibits some indicators, a *capacity* for "consciousness." Yet many key elements are missing. And that is exactly where we were with Debbie and with Kate before her. Some functions of normal consciousness—speech perception for Debbie, face perception for Kate—were present. But there wasn't enough to conclude that either was conscious. *Frustrating*, to say the least.

Questions about when consciousness first begins affect us all in one way or another. Consider some of the concerns that are often raised about abortion and the right to life. We were all fetuses once, subject to the vagaries of lawmakers who often appear more easily swayed by political lobbyists and religious zealots than by scientific evidence.

If you think life begins at the moment of conception and/or believe in the sanctity of all human life, then the question of when consciousness emerges is probably a moot point. But for the rest of us, much of the intellectual baggage surrounding the debate about abortion concerns the possibility that a fetus, at a particular stage of development, might be conscious and therefore, in some sense, may "know" its fate. A related concern is that if a fetus is conscious, then it may have the capacity to "feel" pain. Feeling pain is an experience; it is not a physical property of the outside world, such as temperature, but a personal experience that each of us has in response to a common trigger.

When a thorn pricks your finger, or you realize that you have just put your hand on a hot plate, your experience will be different from mine. It will depend on your previous experience of pain, your state of mind, and the internal chemical milieu of your body and brain. Pain is a conscious experience, and to experience pain, we must be conscious. If that were not the case, then anesthetic drugs such as propofol would not allow us to withstand the agony of surgical pain. The trigger (in this case, the surgeon's knife) hasn't changed, but the conscious experience, thankfully, has.

What we do know is that the fetal brain does not even begin to develop until three to four weeks after conception, so the most basic building blocks of pain perception, the scaffolds of consciousness, do not exist before then. The major divisions of the adult brain emerge at four to eight weeks into pregnancy, but only after about eight weeks does the cerebral cortex separate into two distinct hemispheres. At twelve weeks, rudimentary neuronal connections are emerging between different parts of the brain, but these are not sufficient to support conscious experience.

As Daniel Bor argued in his excellent 2012 book, *The Ravenous Brain*, the areas of the brain that need to be intact, functional, and able to communicate with one another for conscious awareness to occur do not get properly laid out until about twenty-nine weeks into the pregnancy, and it's another month before they are communicating effectively. On the basis of the science then, it's highly unlikely that consciousness in any form, including the ability to experience pain, emerges before about thirty-three weeks after conception.

Detractors point out that a fetus as young as sixteen weeks responds to low-frequency sounds and light. Indeed, by nineteen weeks a fetus may flinch or withdraw a limb in response to a painful stimulus. These are persuasive signs, and it's understandable why they are often taken as evidence for emerging consciousness. However, as Daniel says in his book, these responses are generated by the most primitive parts of the brain, which are unconnected to consciousness, and do not therefore in any way imply that the fetus is aware. What we are witnessing are early reflexes, probably controlled entirely by a primitive brain stem and spinal cord, to a set of physical circumstances and conditions. Someone with a religious orientation might point out—with some justification—that this view still doesn't explain *what makes consciousness happen.* It's almost as if a mysterious switch were flipped on. Precisely because we don't fully understand how or when that switch is turned on, God's will—his "grand design"—is often invoked as an explanation.

As a scientist who has devoted much of my life to understanding whether consciousness exists in people in extremis, I think such arguments are entirely spurious. That we don't yet know what makes consciousness happen has no bearing on whether it can be physically explained. Indeed, I have no doubt that these things will be understood and explained in *the near future*, just as many of the other great mysteries of the universe have in recent years been explained by physics. As scientists, we collect data, we generate hypotheses, and we test those hypotheses. Sometimes we solve the problem and explain something new, and sometimes we don't. But whether or not we solve the problem today has no bearing on whether it is solvable. Falling back on metaphysical explanations just because we haven't yet found the physical answers is antiscientific, illogical, and, to my mind, irrational. After all, if we did that all the time, we'd still be trying to avoid sailing off the edge of our flat earth!

≈

Just as we were wrestling in Cambridge with the question of whether Debbie was conscious and looking hard at when consciousness begins, it appeared that an entire country on the other side of the Atlantic was going to war over when consciousness ends. The gray zone was suddenly the lead story on the US evening news, and word quickly spread to our side of the pond. Somehow, a perfect storm erupted: the right patient, the right family, the right disagreement, and the right amount of public interest in an issue that had, until then, garnered precious little media attention. The right-to-life and the right-to-die movements faced off over one woman who had been declared vegetative and lay in her hospital bed, apparently unaware that half a nation was going to bat for her. Theresa Marie "Terri" Schiavo had had a cardiac arrest at her Florida home in 1990 and sustained massive brain damage from prolonged oxygen deprivation. In 1998, her husband, Michael, petitioned the Florida courts to remove her feeding tube so that she would be allowed to die. Terri's parents, Robert and Mary Schindler, opposed him, arguing that their daughter was conscious.

Cambridge looked on agog. Book deals were signed, documentaries shot, the family appeared on reality TV, lawsuits were launched, protesters for both the right-to-life and right-to-die movements took to the streets, the press seethed. To us Brits it was simply absurd. You might well imagine the conversation over tea and croquet.

"Well, at least the president's not involved."

"Oops! The president is involved."

With the Monica Lewinsky–Bill Clinton debacle and the O. J. Simpson trial recently behind us, we had warmed up to the idea that the American legal system is unpredictable at best and occasionally absurd.

As if to accentuate the contrast, Britain was recovering from its own Schiavo debacle, which lacked the circuslike atmosphere of Florida but was nonetheless heart wrenching. Anthony Bland, a twenty-two-year-old supporter of the Liverpool soccer team, was injured in the Hillsborough stadium disaster—a stampede that had killed ninety-six people in 1989. Bland's case preoccupied the country for months and the courts for years. Fans blamed the police; the police blamed the fans. Bland suffered severe brain damage that left him in a vegetative state. The hospital, with the support of his parents, applied for a court order that would allow him to "die with dignity."

The judge, Sir Stephen Brown, ruled, for the first time in an English court, that artificial feeding through a tube is medical treatment and that to discontinue treatment would be in accordance with good medical practice. Opposition was immediate, but of a very British sort. The lawyer appointed by the Official Solicitor to act on Bland's behalf argued that to withdraw food from him would be tantamount to murder and appealed the decision. The appeal was rejected by the House of Lords.

In 1993, Bland became the first patient in English legal history to be allowed to die by the courts through the withdrawal of life-prolonging treatment, including food and water. There was relatively little opposition, not much fuss, just a rather sober treatment by the media, who noted that times had now changed and in cases where there "was no hope" patients should be allowed to exercise their right to die.

It was a peculiarly British way of doing things. Respectful, mournful, and stoic, with no more than a modest departure from standard protocol. In April 1994, pro-life campaigner Father James Morrow did attempt to get the doctor who withdrew food and drugs from Anthony Bland charged with murder, but the petition was quickly rejected by the High Court.

This was not at all the atmosphere or attitude in the United States, where the party was in full swing. In 2003, "Terri's Law" passed in Florida, which gave Governor Jeb Bush the authority to intervene in the case. Bush immediately ordered the reinsertion of Schiavo's feeding tube, which had been removed a week earlier.

The Schindlers created more publicity by lobbying to keep their daughter alive. They selected a notable pro-life activist, Randall Terry, as their spokesman and continued to pursue their available legal options. The madness escalated. The case drew the attention of everyone with a microphone and a mouth.

Finally, in 2005, a court allowed Schiavo's husband, Michael, to pull the plug for good. In all, the case involved fourteen appeals and numerous motions, petitions, and hearings in the Florida courts; five suits in federal district court; extensive political intervention by the Florida state legislature, Governor Jeb Bush, the US Congress, and President George W. Bush; and four denials of certiorari from the Supreme Court of the United States. As legal expert David Garrow put it in the *Baltimore Sun*, "The most-reviewed and the most-litigated death in American history" was over.

Schiavo's autopsy revealed widespread brain damage, with profound shrinkage to key cortical regions. After an injury or a prolonged period without oxygen, brain cells often die off and never get replaced. This is called apoptosis, a common pattern in vegetative-state patients. Damage to parts of Schiavo's cortex critical for higher aspects of cognition—thinking, planning, understanding, and making decisions—makes it quite clear that she retained no semblance of awareness. The basic building blocks of cognition, the scaffolds upon which our consciousness is supported, had been demolished.

Understanding whether Terri Schiavo was conscious is not like understanding whether a one-month-old is conscious. While one-month-olds' behavior is confusing, they do have the neural machinery that is required for consciousness to exist whether it is there or not. Schiavo had neither the machinery nor the potential. She was not in the gray zone. The person who was born

Theresa Marie Schindler in Montgomery County, Pennsylvania, a shy woman who had married her first love, Michael Schiavo, did not exist anymore and never would. What had replaced that person? It was hard to say. What was indisputably clear was that Terri Schiavo was long gone.

The Schiavo case crystallized public awareness of the gray zone. It brought brain injury and science into the courtroom for the first time on a mass scale in an explosive frisson of science, law, philosophy, medicine, ethics, and religion. I realized that by investigating the gray zone, we were really investigating what it means to be alive. We were exploring the border between life and death. We were right at the nexus of trying to figure out the difference between a body and a person, the difference between a brain and a mind. As the great Francis Crick, a physicist and molecular biologist, wrote in his seminal 1994 book, *The Astonishing Hypothesis*, "You, your joys and your sorrows, your memories and your ambitions, your sense of personal identity and free will, are in fact no more than the behavior of a vast assembly of nerve cells and their associated molecules." Only a few years later, we were beginning to uncover how that three-pound lump of gray and white matter in our heads generates every thought, feeling, plan, intention, and experience we ever have.

CHAPTER SIX

PSYCHOBABBLE

≈

> The limits of my language mean the limits of my world.
>
> —Ludwig Wittgenstein

As the "right to live" versus "the right to die" divided two nations, we were busy trying to build a body of evidence that would help us to understand the minds of people like Terri Schiavo and Anthony Bland. We needed more evidence; more reliable evidence. Evidence that was completely incontrovertible. The Schiavo circus had made that abundantly clear. I was convinced that the stakes of what we were doing were even higher—if we could parse what enabled Debbie's and Kate's brains to respond to our "stimulation," we would be on the road to cracking the code of consciousness itself.

Our next step was to design an experiment that would allow us to conclude that a patient like Debbie or Kate had the capacity to *understand* language. We knew their brains could process speech, but did *they*, the people inside, have any sense of what that speech actually meant?

Ingrid Johnsrude and her colleagues Jenni Rodd and Matt

Davis were working on exactly this problem at the Unit, pinpointing which parts of our brains are responsible for understanding spoken language. The reasoning behind one particular experiment was elegant and—in true Unit tradition—a little quirky. If they submerged speech in a sea of static noise, then those parts of the brain responsible for understanding language would have to work harder to extract meaning from what was being heard and thereby make themselves known on a PET scan. They were constructing experiments that were similar to turning the frequency dial on a car radio, looking for a decent signal. Sometimes you'll chance upon a station where people are talking about something of great interest to you, but the reception is terrible and you can hardly make out what is being said. The lure of the subject matter keeps you listening, but you have to strain to decipher the content of the conversation from the background noise.

Ingrid and her colleagues created a situation almost exactly like that in the PET scanner for a group of healthy volunteers. They were played sentences that varied in terms of their "intelligibility." The amount of static noise, relative to clear speech, was adjusted so that some of the sentences were easily understood, some could be deciphered with a bit of extra effort, and some were almost incomprehensible. As the sentences became increasingly difficult to decipher, activity increased in an area of the temporal cortex on the left side of the brain. The more difficult the sentences were to understand, the harder this region of the brain had to work, and this showed up on the PET scan as more and more radioactive blood rushed to that region to replace used-up energy.

My psycholinguist friends had found a way in—a way to distinguish between a brain that was *understanding* speech and one that was merely *experiencing* it. Could this be the answer? The key to unlocking the conscious mind? We needed another patient to answer that question.

≈

In June 2003, Kevin, a fifty-three-year-old bus driver from Cambridge, collapsed with a severe headache and quickly became drowsy. The next day he was unresponsive, paralyzed down one side with strange, uncontrollable eye movements. After he was admitted to Addenbrooke's, an MRI scan revealed that he had experienced a massive stroke in his brain stem and thalamus—the ultimate "double whammy" for consciousness.

As we've seen, many of the brain's most essential functions, including sleeping and waking cycles, heart rate, breathing, and consciousness, depend on the brain stem. The brain stem also sends a multitude of sensory signals about hearing, taste, and the sense of touch and pain to the thalamus. The thalamus is a central relay station, or a hub, connecting multiple brain areas in an incredibly complicated network of communicating neurons. The relationship between the brain stem and the thalamus is crucial for holding it all together, maintaining consciousness, and keeping us alive. It's the be-all and end-all.

After his admission to Addenbrooke's, Kevin's level of wakefulness fluctuated, but then stabilized into deep unresponsiveness. Three weeks of follow-up assessments revealed no change in his condition and he was declared vegetative. Four months after he collapsed, in October 2003, his condition was considered stable enough for us to scan him, and we decided to give the new test Ingrid and her colleagues had developed a go. We'd scan Kevin as we played him the recorded sentences submerged in static and try to figure out whether he could understand them. It seemed like a long shot but worth a try.

As Kevin's scan began, I wondered if, after Kate and Debbie, we could get lucky a third time. Amazingly, we did! We saw a strong response in Kevin's brain areas specifically devoted to processing speech sounds. This was exciting but not new; it was almost exactly the same thing we'd seen in Debbie's brain when we'd played her single words, with signal-correlated noise for comparison. But it did tell us that Kevin, like Debbie before him, was still processing speech as he had done before his injury.

In Debbie's case, that's where the case had run cold. We could ask no more because we had only tested her with speech and non-speech sounds, nothing in between. The question of whether her brain could *understand* speech had therefore remained unresolved. With Kevin we had more. We carefully compared what happened in his brain when we played him sentences that were easy to understand with ones that could only be understood with a bit of extra effort, and with sentences that were difficult to decipher.

Incredibly, this revealed two sparks of brain activity in the upper and middle ridges of the left temporal lobe of Kevin's brain. These were the same parts of the brain that were activated when healthy participants had to try that little bit harder to make sense of the sentences submerged in static. Put simply, language comprehension is strongly related to brain activity in the left temporal lobe in healthy participants. In Kevin, supposedly in a vegetative state for four months, activity in this same part of the brain changed as we made a series of sentences increasingly incomprehensible. Surely this was key evidence that Kevin's brain wasn't just hearing speech—his brain *understood* it!

≈

Nine months after we first scanned Kevin nothing had changed. He was still in a vegetative state, still entirely physically non-responsive, and still in the hospital. We decided to rescan him. The results were identical. His brain lit up when we played him the same sentences as before; the activity was stronger when those sentences were submerged in static that made them more difficult to understand. The areas of his brain that lit up with each scan were almost identical to those that we had seen nine months earlier. We had replicated our findings. There could be little doubt that Kevin's brain was processing *meaning*.

While it was satisfying to replicate our study, it was also frustrating. What I really wanted to know was what it was like to be Kevin, and whether we could do anything to alleviate any suffering that he might be experiencing. Did he feel the same raging thirst that Kate had felt? Had he tried to end his life by holding his breath? Was he listening to every conversation, or had he left this world and detached himself from the nightmare that his life had become? Was he aware that we had scanned him? Did he know that we had tried to make contact? Did he even care?

These questions were tantalizing, but I knew that to answer them we had to remain focused, proceed step by step, take each piece of scientific data, scrutinize it, then use it to build up a picture of what was going on in Kevin's world. If, indeed, he had a world at all.

≈

In both Kevin's and Debbie's cases, we were still trying to understand the ways in which language and consciousness were related. We'd moved forward, but many of the thorny old questions about consciousness persisted. Kevin's brain could understand the meaning of sentences. Did it mean that when he heard a sentence such as "The man drove to work in his new car," Kevin experienced the event in his mind's eye, a fully fleshed-out stream of imagery that he could reflect on and even embellish? Or was his response at a lower, more automatic level; not so much an experience that could be reflected upon, but a simpler association between words and their meanings such that the sentence conjured up an image of a man and a car but little else? *Man*, *work*, and *car* are all common nouns that could register because of their familiarity in the machinery of the brain, yet they could have been for Kevin (and others like him) devoid of the detail or imagery that is part and parcel of our fully conscious experiences.

Many of our most complex brain processes, even our ability to understand speech, can go on when we are less than fully aware. If you're asleep—perhaps not deeply asleep but asleep nonetheless—and someone close by utters your name, you might wake up. Yet if someone close by utters someone else's name, particularly that of someone who is of no importance to you, you may doze right through it.

That you respond differently to those two situations confirms that your brain, while in reduced awareness, is monitoring and making decisions about the contents of speech in your vicinity. It can't be that somehow your brain "hears" your own name, but "doesn't hear" other names, because if any name went "unheard," then your brain would have no idea whether it was your name or not. The brain must be registering *all* names.

Take the logic a step further. As you sleep, your brain must be monitoring and processing all of the speech around you, indeed, *all of the sound around you*, just to be able to "decide" whether it's your name, someone else's name, no name at all, or the sound of a distant lawn mower. Through most of this you are asleep and unaware of what is happening around you and the way your brain is processing it. This doesn't just apply to humans. Watch your cat or dog sleep soundly through a loud but familiar sound (such as a lawn mower), but open an eye when they hear something quieter, but a whole lot more interesting—a mouse scratching in the cupboard! It's not hard to understand why this is so; it's crucial for survival and has probably been part of our repertoire of attentional capabilities for millennia. We all need to be awakened when something that is potentially dangerous (or edible) makes a sound. But imagine if every sound had the same effect—we'd be up and down all night!

How then should we interpret the activity inside Kevin's head? Was it conscious, or was it just his brain doing its thing while he, Kevin the *person*, remained unaware?

There was no clear answer. We had to dig a little deeper. I hoped Kevin's brain activity was a sign, a tiny message telling us that he was still in there, wanting to get out, waiting for us to find him and release him from what I could only imagine was his tormented existence. But another part of me shivered at the thought. I hated the possibility that Kevin might be in there, aware that we'd scanned him, but equally aware that now we were stuck wondering what his brain activity actually meant. After all, if Kevin *was* conscious, then he would have been a party to every one of our conversations in his presence, he would know that we had been trying to make contact with him when we scanned him, and he would know that we had no idea how to interpret the result. Like a castaway stranded on a desert island, were we the ship that had just passed by in the far distance, leaving him frustrated and confused? Had we made his situation worse by adding to his misery? I tried not to think about it.

Whatever Kevin was experiencing, meeting him and making contact with his brain left me dwelling once again on Maureen's predicament and wondering whether any parallels were likely between their two situations. The origins of their brain injuries were certainly very different, but where these had taken them—to wakeful unresponsiveness—was more or less identical. If Kevin was in there, could Maureen be in there too?

≈

Then, everything changed.

After many months of tweaking and cajoling, the Wolfson finally acquired a functional magnetic resonance imaging scanner, or fMRI. This remarkable technology, developed for use in humans in the early 1990s, opened up a whole new world of possibilities and revolutionized the development of gray-zone science.

fMRI uses an entirely different technological approach to brain imaging than PET, yet the results—detecting brain activity associated with thoughts, feelings, and intentions—are much the same. Blood that is carrying oxygen to the brain behaves differently in a magnetic field than blood that has already delivered its oxygen. In other words, oxygenated blood and deoxygenated blood have different magnetic properties. More active areas of the brain receive more oxygenated blood, and the fMRI scanner can detect this and pinpoint where the activity is occurring. Unlike PET, fMRI has no "radiation burden." In fact, fMRI has no harmful effects at all, so patients can be scanned again and again. When positive results start to come in, you can keep going to try to work out exactly what's going on. The case never has to run cold.

fMRI has other advantages that are even more significant. Brain activity can be monitored second by second instead of over a period of a few minutes, as was always the case with PET. This has far-reaching consequences. One of the most important is for studies that involve spoken language. The brain processes that allow us to understand language operate over seconds, not minutes.

Reading and understanding this page of text usually takes you about a minute—about the length of a PET scan. But by the time you get to the end of the page, your brain has decoded and understood a number of different sentences. You don't wait until the end of the page to digest its contents. In fact, you couldn't even if you wanted to.

Understanding language is ongoing, and your brain deconstructs a page of text into its overall meaning one piece at a time, sentence by sentence. Actually, understanding meaning occurs at an even lower level than that, as we shall shortly see. For now it's sufficient to say that the size of a chunk of information that can be investigated with fMRI—its "temporal resolution"—is sufficient to unpack how we process single sentences. The temporal resolution of PET scanning was minutes, rather than seconds. You could only examine how the brain responds to a whole page of text, while fMRI allows you to examine how each sentence is processed and understood.

This was a crucial development because our problem with Kevin was nailing down exactly what he could understand. Perhaps it was just big ideas, general themes, a rough gist of what was going on. Or could he pick out the contents of spoken language sentence by sentence, word by word, just like you and me?

≈

Similar to reading, understanding clear speech in our native language is ordinarily so effortless that we're unaware how complicated it really is. Not only do we have to identify all the individual words, we must also retrieve the meanings of these words and combine them appropriately to understand a sentence.

A huge fly in the ointment is the ambiguity of many words in English (about 80 percent). Homonyms have two meanings with the same spelling and pronunciation (*bark*). Homophones have two meanings that have the same pronunciation but different spelling (*knight*, *night*). In the sentence "The boy was frightened by the loud bark," you have to work out that the ambiguous word *bark* refers to the sound made by a dog and not to the outer covering of a tree. Your brain does that by using the context provided by the rest of the sentence. It's possible with fMRI to see how a single sentence such as "The boy was frightened by the loud bark" is decoded into its correct meaning by our brains in milliseconds.

Ingrid Johnsrude and her colleagues were using semantic ambiguity to try to work out how the healthy brain understands spoken language. They had carried out an fMRI study in which healthy participants lying in the scanner had heard sentences containing several words that had more than one meaning: "The shell was fired toward the tank" (*shell*, *fired*, and *tank* all have alternate meanings). The participants were also played sentences that contained no ambiguous words: "Her secrets were written in her diary." While the two types of sentence were well matched in all sorts of important psycholinguistic ways, the theory was that those containing ambiguous words would require additional brain processing to identify and select their contextually appropriate meanings. Sure enough, the sentences with ambiguous words produced increases in brain activity in the left temporal cortex and in the lower part of both frontal lobes, meaning that these two regions are important for understanding the meaning of spoken sentences.

This was crucially important information for us as we pondered the results of Kevin's PET scans and what his understanding of language might actually be. The simple task in which participants lying in the scanner heard two different types of sentences appeared able to reveal whether someone's brain could decide between two possible meanings of an ambiguous word, by relating that word to the context (or "meaning") of the rest of the sentence. Surely this was language comprehension at the highest level? What more is there to understanding language? Does it get any harder than that? We were no longer talking about language comprehension in the vague, half-baked sense of a general, perhaps automatic, association between a word and its meaning (I know that a "dog" is some kind of "animal"). Now we were talking about whole sentences—whole *ambiguous* sentences—being understood in a way that can only mean that the multiple meanings of each word had been retrieved from memory, and then the appropriate meaning selected based on the relationship of each word to contextual information provided by the rest of the sentence.

What we were starting to realize is that understanding language may be the key to consciousness—not in the sense that language *is* consciousness, but that if people can be shown to understand language at its most complex, then they are likely conscious. Philosophers might argue that voice-to-text translators, such as Siri, *understand* language in some sense, yet they would all probably agree that Siri and her peers are not conscious. However, it is in situations precisely like the one described above, with semantic ambiguity, that machines (and not humans) come unstuck. Neil Armstrong and Buzz Aldrin walked on the moon almost fifty years ago, yet the best minds on the planet still seem incapable of building a machine that will understand human speech without error.

Why? Part of the problem is that human speech is riddled with ambiguity even when the individual words are not ambiguous. Consider the sentence "He fed her cat food." Did he feed food to his female friend's cat, or did he feed cat food to his female friend? It's impossible to know based on that one simple sentence because the sentence is ambiguous. Our brains usually cope with this ambiguity by considering context. Were we talking about his female friend's cat when the sentence was uttered? Or his female friend's strange eating habits? How can machines or pieces of software tell the difference? They can't (or at least they mostly can't), because unlike you they are not "aware" of everything that has happened to you that minute, earlier that day, last week, or at any other point in your lifetime—the information that you have that provides context and allows you to understand which of the two explanations for the sentence "He fed her cat food" is applicable now.

It bears repeating that Ingrid and her colleagues had shown us that two brain areas, one on the left side and toward the back and the bottom of the temporal cortex and another toward the lower part of the frontal lobes, are important for understanding the meaning of spoken sentences. Where there's ambiguity, these areas try to solve it. But it gets even more complicated than that. The brain's memory network is also crucial for understanding spoken language. If we remember that our female friend does not have a cat, then the interpretation of "He fed her cat food" that has her eating a can of Whiskas becomes a bit more likely. Yet we know, from memory, that people generally do not eat cat food. Cats eat cat food. So gradually all of these brain processes work together to solve the problem of language ambiguity.

Herein lies the connection between language and consciousness. Because so many complex cognitive processes are involved in understanding the meaning of language, processes that involve word disambiguation, decoding context, retrieval of information from long-term memory, and appreciation of social norms (few of us eat cat food), it was starting to seem that if a brain can be shown to be performing all of these processes effectively, it was simply implausible that it was unconscious. Through language, we were gradually working out what the building blocks of human consciousness must be, one brick at a time.

≈

Kevin became the first patient that we put into an fMRI scanner, the incredible new technology that was to play such an important role in the development of gray-zone science. His feet, in socks, protruded from the scanner's long tunnel. The machine clicked into action with a whir and a dull thud. A burst of radio waves was released, and the unmistakable (and extremely loud) *pip . . . pip . . . pip* of the fMRI scan began.

Kevin was taking part, wittingly or unwittingly, in advancing gray-zone science, adding to our understanding of what it means to be conscious. Yet taking part in our *experiment* would probably have no benefit to him personally. This scan was an important part of a jigsaw puzzle, but we were still a long way from being able to help people. I took heart that Kevin was one of many pieces that were rapidly coming together and the prospect of clinical benefit for other patients, those who would follow Kevin into the gray zone, was imminent.

When we played Kevin the sentences with ambiguous words, his temporal lobe lit up in exactly the same way that it had in the healthy volunteers. We knew from previous studies that the focused left-hemisphere activity, toward the bottom and near the back of the brain, was important for processing meaning. Despite his vegetative-state diagnosis, Kevin's brain was still activating, selecting, and integrating contextually appropriate word meanings to understand complex sentences containing ambiguous words.

No experiment like this had ever before been conducted—a highly sophisticated set of psycholinguistic sentences had elicited extremely subtle changes in areas of the brain that control the most complex aspects of language comprehension. Kevin's brain, it seemed, was still processing complex sentences to create some sort of meaning.

≈

Several months after Kevin's fMRI scan, I excitedly presented our results to a distinguished gathering of clinical and nursing staff in Cambridge. I felt that we had learned something new about

Kevin and about what patients like him were capable of. We were pushing the boundaries. But the reaction I received from my audience was simultaneously crushing and illuminating. What we had shown—Kevin's brain responding to highly sophisticated, ambiguous sentences—wasn't enough. The audience wanted me to be able to put my hand on my heart and say, "The results of the scans confirm that Kevin is definitely conscious." However complex the psychological stimuli, however advanced the technology, however smart we thought we were—until we could provide irrefutable evidence that Kevin was conscious, no one would believe that he was. Or even that he might be.

≈

I don't know whether it was my scientific frustration with Kevin and where his scans had led us, but in 2004 I decided that I needed a break. The previous year I had been invited to Sydney, Australia, to give a keynote lecture on my work on frontal-lobe function and Parkinson's disease, and I had made some new friends in the psychiatric community at the University of New South Wales. They had recently acquired a new fMRI scanner and extended an invitation for me to return, for a longer stay, to assist them in getting their imaging program launched.

I seized the opportunity to get away and went down under for four glorious months, renting an apartment on Coogee Beach, just a couple of bays south of Bondi, with its golden sand, beautiful people, and perpetual sunshine—about as close to paradise as you could get for a Brit. I spent mornings on the beach or walking the beautiful cliff path. I was alone. And I had lots of time to think.

It had been eight years since Maureen's accident. There had been Kate, barely a year afterward. Then Debbie and Kevin. The Schiavo debacle was coming to a close. Within months she would be dead. My scientific interests were gradually shifting from the work that I had spent most of my career pursuing—the functions of the frontal lobes and their relevance to conditions such as Parkinson's disease—to the emerging field of consciousness in patients trapped in the gray zone.

This new direction was impossible to ignore. It was exciting, invigorating, and—in a strange, scientific way—seductive. It was brain imaging with a *purpose*. No longer just science for science's sake. A scientific journey with the clear prospect of an outcome that would benefit real people with real problems. Real people such as Maureen. Quite how we'd get there I did not know. Every experiment generated as many questions as it answered, but every new question was as intriguing as the last.

The only problem was, I didn't know where to take it. What was the next step? What was the next question we needed to ask to advance our understanding? I was stuck. And then it occurred to me—the answer was at my fingertips. The two seemingly unconnected strands of my research were not so unrelated after all. In fact, they were very closely related indeed. The next line of investigation had been staring me in the face. I just hadn't seen it.

CHAPTER SEVEN

THE WORLD AS WILL

≈

> Whatever torch we kindle, and whatever space it may illuminate, our horizon will always remain encircled by the depth of night.
>
> —Arthur Schopenhauer

I last received news about Kevin in 2005, more than two years after his stroke. By then he was stable and living in a residential-care facility, but his vegetative-state diagnosis was unchanged. I wondered whether he knew we had tried to reach him. The staff at the care home were aware of our findings, but would they make a difference in Kevin's life? Would he be treated differently? Would the staff talk to him because he might be able to understand them? Would they read to him? I would probably never know. It was frustrating, but there was nothing I could do.

Around the same time that we scanned Kevin, I had been working on an fMRI project with Anja Dove, one of my postdocs, investigating how our frontal lobes contribute to memory. Our intuition said the frontal lobes are important for those occasions when we specifically set out to lay down a memory, when we tell ourselves that we *need* to remember something. They are not crucial for what one might call "automatic" memories, those details and facts that you effortlessly acquire as you go through life whether you want to or not: what your car looks like or how to find the bathroom in your own home. Your frontal lobes come into play on those occasions when you actively set out to remember a telephone number, an address, or a shopping list that is too short to bother with writing down. The distinction was important for the line of investigation that was developing in my mind: to show that consciousness existed in at least some people who outwardly appear to be in a vegetative state—people whom many insisted exhibited automatic, nonconscious responses in the scanner to the stimuli we presented.

As I watched the surf on Coogee Beach, various strands of thought started to coalesce and take shape. Then, in one of those inspired moments that comes along only when you least expect it, I realized that intention and consciousness are inextricably linked; if we could demonstrate one, then we could assume the other. And *intention* was exactly the form of cognition that we were already exploring through our frontal-lobe memory experiments. To understand this requires some further explanation.

Imagine that you are wandering through an art gallery. In an hour or so, you see hundreds of paintings, some unique and distinct, some similar in terms of color, subject matter, or style. Imagine also that you make no particular effort to remember any of them. Much later, if you revisit that same art gallery, you'll probably recognize some paintings but not others. Some may look familiar, but you can't be absolutely sure whether you've seen them before. Although you might think you recognize some of the paintings, you're actually confusing them with other paintings that you saw that were similar in some way.

This is how most memories work; loads of information is out there in the world to remember, and real life is not like a memory test, so we don't go around trying to remember what we experience in an effortful, conscious way. We just live our experience. Some of it sticks and some of it doesn't. Generally, what sticks is unique and distinct, and what doesn't is the information that's similar to other information we've experienced and is therefore more easily confused.

That's not to say that we walk around in a daze—well, at least not most of the time. We generally have an "attentional spotlight" (as some cognitive neuroscientists call it). Things within that spotlight stand a good chance of being remembered, whether we like it or not. When we bring our attention to bear on something, it forms a representation in our brain—clusters of neurons that fire in response to its size, shape, how it sounds, looks, and feels, what it is similar to, whether we have seen it before. Every aspect of something in our attentional spotlight, from its physical properties to its location and its relevance to other objects present at the same time and present in our heads (such as previous memories), gets "represented" by neurons firing. That is the physiological basis of attention—a remapping of something in the physical world, such as an object that you are looking at, onto a network of firing neurons in the brain. Because that particular network of neurons fires together, the chance of their being laid down as a memory—an ongoing, stable representation that can be retrieved at a later date—is heightened. To paraphrase the famous twentieth-century Canadian neuropsychologist Donald Hebb: "Neurons that fire together, wire together." What Hebb meant is that every experience, thought, feeling, and physical sensation we have triggers thousands of neurons, which form a neural network, or "representation," of that experience. With each repetition of that experience, the connections between those neurons get stronger and the "representation" gets more and more "hardwired" as a "memory" in our brains.

This type of memory—the memories over and above those

you actually set out to remember (such as your times tables)—is carried out by the temporal lobes of the brain. It's all relatively automatic and outside our conscious control. Psychologists refer to it as recognition memory because often the only time that we become aware of it is when we spontaneously "recognize" something we've experienced before. You don't need your frontal lobes for recognition memory. Back in my Maudsley days with Maureen, I showed that patients who had sustained massive damage to the frontal part of their brains were still able to recognize a picture they had seen before, even if they'd only glimpsed it briefly. On the other hand, our patients with temporal-lobe surgery had real problems spotting a picture they'd been shown just a few seconds earlier. The frontal lobes only spring into action when we actually want to remember something specific, when we have a conscious desire or thought to commit something to memory.

Why we have these two different ways of laying down memories is unclear, but it is tremendously powerful and certainly has a lot to do with consciousness. If all we were able to remember were the things we intentionally set out to remember, we'd be in a lot of trouble most of the time. Imagine meeting your mother-in-law for the first time and forgetting to make a special point of remembering her face. It would be embarrassing when you failed to recognize her the following day. It's great that our brains remember things like that automatically because then we don't have to remember to do it ourselves. It's efficient because much of what we remember, even much of what we *need* to remember, doesn't need to be consciously and meticulously learned. It's good enough just to know that when you see your mother-in-law again, you'll recognize her.

On the other hand, you don't want your entire memory to work on autopilot all the time—you want to have some capacity to *decide* what is most important to remember. If you are introduced to your mother-in-law at the same time that you are introduced to a whole gaggle of aunts and distant cousins, you need to focus on your mother-in-law's name and remember that, for there's little doubt that the penalty for forgetting that name in the future will be highest. A few moments in your attentional spotlight isn't going to cut it with the mother-in-law. You need to steal yourself away for a moment of conscious thought, activate your frontal-lobe memory system, and make a special and intentional effort to remember that one name above all else. This is where consciousness really comes into its own.

The intention, the willful decision to commit something to memory, rather than leaving what is remembered and what is forgotten to the vagaries of your temporal-lobe memory system, is a conscious act. Just like remembering your times tables, remembering your mother-in-law's name will serve you well and is worth investing some conscious energy to do.

On the beach in Coogee, I started to realize that understanding whether a memory is laid down automatically or intentionally might be the key to understanding whether a response in a vegetative-state brain is conscious. If you can show that it is *intentional*, then it is certainly conscious. If, on the other hand, it is automatic, then it may not be.

To illustrate this, imagine yourself back in the art gallery. If you are wandering through the exhibits and you want to be sure to remember one particular painting above all others, you make a conscious decision to remember that painting and you willfully (and knowingly) commit it to memory. Much later, when you revisit the gallery, you will have a good chance of remembering that particular painting and a lower chance of remembering the others. Why? Because you used your frontal lobes to assign special importance to that piece of art and you made an *intentional*, effortful attempt to remember it.

Remembering where you parked your car each day is another great example of how the frontal lobes do their work. In that case, you assign special importance to today's parking space in your *working memory*, holding it there only until it's no longer needed at the end of the day (when you retrieve your car). But it's true for longer-term memories as well: those involved in visiting and then revisiting an art gallery or remembering the name of your mother-in-law. If you want them to, your frontal lobes can strengthen a memory trace and increase the chances of your successfully retrieving it later.

If you are inundated with names of aunts and distant cousins, then you may have to bring in the heavy artillery—one particular area within the middle and top half of the frontal lobes known as the dorsolateral frontal cortex. This area of the brain is good at indexing and cataloging—such as if you've got a whole bunch of names, all competing for attention, but just one or a few that you want to assign special importance to (your mother-in-law's name). It can also do some special functions that make memory retrieval more precise (does she like to be called Jo or Josephine?). And it can override persistent and burned-in memories when necessary (if you had been married to Sally for thirty years, it may require some special effort, and some dedicated input from your dorsolateral frontal cortex, to remember that your current wife's name is Penelope). This seems to be part and parcel of what the frontal lobes have evolved for—to give us that extra level of control, that extra sense of being the one making the decisions, calling the shots, the person, the self, the sense of being *someone*.

It should come as no surprise then that this region of the brain has also been associated with aspects of general intelligence (g) and performance on IQ tests. Our ability to reason, to work our way through complex problems, and to plan ahead all depend on our frontal lobes, and these are essential cognitive abilities that dictate how far we will go in life. For example, achievement in school has repeatedly been shown to be related to scores on tests of g, presumably because our g scores depend on our frontal lobes, which in turn dictate our ability to handle our memories smartly in ways that will prove useful to us in a variety of different situations. Again, learning facts isn't enough—it's what you do with them that counts.

≈

I can talk about the nuances in the relationship between how the frontal cortex and the temporal lobes handle memory now, but their interactions were not so clear in 2004 when Anja and I were working on the problem. In true Unit fashion, we "mocked up" an art gallery in the fMRI scanner to test the hypothesis. We showed a group of healthy participants hundreds of obscure paintings that we could be reasonably sure that the participants would not have seen before (and so would not remember from a previous occasion) while we scanned them. Every so often during the scan, we signaled to our participants to make a special effort to remember the next piece. Otherwise there was no such signal or special instruction.

Our hypothesis was spot on. Looking at art with no explicit instructions produced an increase in temporal-lobe but not frontal-cortex activity. Some of these paintings were recalled; others not. When instructions changed to encourage the volunteers to remember a specific painting, we saw increased activity in the frontal lobe, just as we had predicted, with no additional increased activity in the temporal lobe.

More important, after the scan, these particular pieces of art were remembered much better than the rest. This was interesting in and of itself and made a modest impact on the scientific literature on frontal-lobe function when Anja and I published it in the journal *Neuroimage* two years later. But sitting on the beach in Sydney in 2004, I already knew the results, and with Kevin in mind they were starting to take on a whole different kind of significance.

I realized that because the only difference between the conditions that produced frontal-lobe brain activity and those that did not was in the instruction given *prior* to each painting, the brain activity observed must reflect the *intentions* of the volunteer (which were based on the remembered instruction) rather than some altered property of the outside world. That is to say, the paintings that the participants were told to remember (and subsequently remembered better) and those that they were given no instructions about had no physical differences. They weren't *easier* to remember. The only difference was what the participants did when they saw the paintings (that is, try to remember them), and that was based on their conscious intention or *will*.

You may think that I am being disingenuous and that the decision to remember, or not remember, is the result of the instruction that they were given. That is true, but only in part—there is more to it than that.

To return to the art gallery—I have instructed you to select one particular painting, any one you choose, to remember especially well. I have given you an explicit instruction, just as in the experiment that Anja and I conducted in the fMRI scanner. But will you *act* on that instruction? Will you make a special effort to identify one particular painting and remember it? You might not for all sorts of reasons. You might lose yourself in aesthetic reverie and leave the art gallery with no one piece of art given special attention. Or you may just decide to disobey me. I issued an instruction, but you choose to ignore it. It would be pretty easy to wander around an art gallery making no particular effort to remember any one painting, even if you've been previously instructed to do exactly this. The point is, you can issue instructions to participants in scanners, but whether they are carried out or not depends on their will. Their *conscious* will. They may unconsciously forget to follow the instruction, but if they follow it, then it is a *conscious* act, an intention, an act of subjective will. Just like the decision to make a special effort to remember your mother-in-law's name at the expense of all the great-aunts and distant cousins, this is not something that just happens. *You must decide to do it.*

On the beach in Sydney I realized that the decision to "remember" a painting rather than simply "look at it" is clear evidence for consciousness in the healthy volunteers that Anja and I scanned in our study about how the frontal lobes contribute to memory. At the time, we weren't interested in whether our participants were conscious; they obviously were because they were healthy volunteers. But I started to imagine what it would mean if we saw the same thing in someone such as Kevin. What if we told him to remember just a few examples from a whole series of paintings that we showed him, and for those paintings only, we saw his frontal lobe respond? Wouldn't it be absolute evidence that he was conscious? Why else would Kevin's frontal lobe spring into action, just for those particular paintings, unless he had remembered our instruction and *consciously chosen to act on it*?

I knew that I had stumbled on the answer. We had to make a vegetative-state patient respond to an instruction that required a conscious decision to do so. Not something that was automatic, but something that they could *choose* to do or not to do. If they did it, we would have the proof that we needed to silence our doubters.

I had found a way into the gray zone, a path into that elusive inner space that we had been searching for so determinedly, a way to be sure that a signal from within, if it ever came, would reflect the presence of a living, thinking being—a person with a sense of himself or herself, the world, and his or her place in it. The implications were huge. Evidence of a conscious decision was all we needed to prove that consciousness existed. It was the key to everything. If this experiment worked, if we could find a nonresponsive patient who could make a conscious decision that we could detect with our fMRI scanner, then we would know, beyond any doubt, that the person was conscious. Once we were through that door, the possibilities seemed endless. Might our new keyhole to the other side allow us to make contact with these people? To ask them what it was like in there? Could they tell us what they wanted? Could they tell us what they knew of their fate, how they got there, and that they were aware of the passage of time? Could they express their likes and dislikes and what would make them more comfortable? Could they even tell us whether they wanted to live or die? Getting into the gray zone had once seemed impossible, yet now we were a single experiment away from having to come to terms with what we were going to do once we got in there.

It was time to go home.

CHAPTER EIGHT

TENNIS, ANYONE?

≈

I'll let the racket do the talking.

—John McEnroe

I returned to Cambridge and, in June 2004, traveled by train to Antwerp through the Channel Tunnel to give a lecture at the eighth annual meeting of the Association for the Scientific Study of Consciousness, organized by Steven Laureys.

Arriving at the conference, I found my way to the lecture hall at the University of Antwerp, a steeply pitched, windowless room that held several hundred attendees. When my turn came, I delivered a thirty-minute lecture, enthusiastically describing all three of our key patients and ending with Kevin because he best illustrated where we were, scientifically speaking. Kevin's case was the first evidence we had that an entirely nonresponsive patient's brain could decode the meaning of sentences. But did this mean that Kevin was conscious? I let the question hang. It was the right place to do it. Many members of the audience of philosophers, neuroscientists, anesthetists, and other clinicians who investigated consciousness were beginning to coalesce not around consciousness per se, but what happens when consciousness goes wrong.

The field of "disorders of consciousness" was only just emerging, but its main proponents—Nicholas Schiff, Joe Giacino, and, of course, Steven himself—were all there.

At the conference reception, held in Antwerp's beautiful Brantyser Restaurant, I was distracted by the haunting sounds of a single cello pushing through the din. As we sat down to dinner, the cello player, Melanie Boly, a Belgian neurologist-in-training, sat down beside me. She made an immediate impression: charismatic, brilliant, and by far the fastest talker I have ever met. We discussed music and science. She was keen to increase her psychological expertise, and we agreed a trip to Cambridge was just what she needed and arranged with Steven for her to work in my lab as a visiting scholar in May and June of the following year. Melanie was the perfect candidate to help us push the science forward together, and Steven readily agreed to pay her expenses. The following morning, I boarded the train back to the UK with a renewed sense of optimism. I knew what we had to do; the pieces were all beginning to fall into place.

≈

So as the weather warmed in the spring of 2005, Melanie and I set about trying to work out how to convert what we knew about the frontal lobe and its role in intention and will into a workable solution for identifying consciousness in physically nonresponsive patients. I was obsessed with the idea that we needed to use an "active task"—a task that involved some sort of intentional mental activity on the patient's part. We sat in the Unit's garden on an old wooden bench tossing ideas back and forth. Right in the center of the lawn stood a drooping mulberry tree that provided perfect cover from the early-summer sunshine.

Melanie and I needed a mental task that would keep participants occupied, unassisted and unprompted, for half a minute. The first idea that we came up with was nursery rhymes. Could we get patients to sing a nursery rhyme in their heads that would produce a consistent pattern of brain activity? Nursery rhymes are familiar and relatively easy to sing to yourself for thirty seconds.

Our second idea was to ask participants to imagine the face of someone they loved. Kate's brain had activated strongly in response to photos of her family and friends, and it didn't seem too much of a stretch to think that simply imagining faces of the people we love might produce similarly reliable patterns of brain activity.

Our third idea was to ask patients to imagine moving through a familiar environment, such as their own home. Navigating from one place to another, even knowing exactly where you are at any moment, is a complicated business effortlessly achieved. Your hippocampus, a sea-horse-shaped structure deep in your brain, has specialized neurons known as place cells, which were first discovered in rats in 1971 by neuroscientist John O'Keefe and his colleagues (O'Keefe received the Nobel Prize for this discovery in 2014).

O'Keefe found that place cells in a rat's brain seem to "know" where the animal is in an environment. He also discovered that place cells in different parts of the hippocampus fired at different times depending on where the rat went, and this whole network of firing neurons constructed a mental "map" of the rat's environment. Amazingly, if the rat was moved to a different location, the same place cells fired, but in a different configuration that "mapped" that new area. This work was important, in part because nothing like place cells had ever before been discovered, and in part because it laid the foundation for later research that demonstrated that the hippocampus is the location of the brain's "cognitive map." The function of this map is not only to allow us to navigate our way through the world, but also to act as a sort of framework or scaffold upon which all of our memories and experiences can be hung.

Think about how you navigate through a familiar environment, such as your home, to reach your bedroom. How do you know when you're there? You might think that you know because you recognize what you were expecting to find: the bed, closet, dresser, and so forth. But that can't be the case. If it were, then we'd spend most of our lives ambling around until we stumbled upon the places we want to reach. That's not how we do it. We typically go straight to where we want to be because we have a well-developed mental map of where we are and where we are in relation to where we want to go. Successful navigation is a tight coupling between our memories and being able to tell where we are in our environment.

If you close your eyes and imagine walking through your house to your bedroom, you can get a sense of this mental map. That we are able to do this demonstrates that our brains have a spatial map all laid out. We can refer to it even when we can't see it in reality. Indeed, most of us would have no trouble finding our bedroom in our own house even with our eyes closed. It might be slow, but we'd get there. The hippocampus is the part of the brain that allows us to do this. It literally maps out your environment for you so you know where you are.

In fact, it's a little more complicated than that, and while important, the hippocampus is not the be-all and end-all of creating our mind maps. Close to the hippocampus, in an area of cortex called the parahippocampal gyrus, is a piece of brain tissue that becomes highly active when humans view pictures of places, such as landscapes, city views, or rooms. It activates reliably whenever you think about moving through a familiar environment.

Melanie and I had our three tasks: singing in your head, imagining faces, and spatial navigation. We knew that not every idea we came up with would work out (it rarely does), but we hoped that one or two of them would provide us with what we were looking for—a reliable task that almost everybody could "do in their heads" with the simplest of instructions.

Melanie found twelve willing volunteers and put them through their paces. The results were mixed. The spatial-navigation task worked well—people easily imagined walking through their homes: we saw a flicker of fMRI activity in the parahippocampal gyrus in all but one participant. The nursery-rhymes scans were inconsistent: some people's brains activated; some didn't. And among those that activated, the brain activity was often in completely different places. The scans in which the participants were asked to imagine faces of people they loved were also disappointing, but for a different reason. Although the activity in the brain was fairly consistent from person to person, many participants reported that it was just too hard to do. It wasn't that they couldn't easily imagine the face of a person they loved, but it was impossible to hold that image in mind long enough for us to capture it with our scanner.

One task out of three was usable in patients. It wasn't enough. We needed something else—a killer task that would work in everybody *all the time*. We went back to my office and looked out over the beautiful lawn, pondering. Melanie mentioned that she had been reviewing the scientific literature on mental imagery, and it seemed that complex tasks worked better than simple ones. What we needed was something complex that was easy to imagine. And then I hit on it. As Melanie recalled recently, I suddenly yelled, "What about *tennis*!?"

Perhaps I struck upon this idea because it was late June and Wimbledon was in full swing. Every summer, between bouts of tea on the croquet lawn, the Unit tuned in to the on-court action in South London, just seventy-three miles away. Or maybe the tennis idea was just dumb luck. But that was the pivotal moment, the turning point, the *nexus* where everything changed. The culmination of almost ten years of thought that would finally allow us to unlock the minds of patients like Kate, Debbie, and Kevin.

Melanie and I laughed at the thought of getting vegetative people to imagine playing tennis in the scanner. It felt like an absurd idea, even by Unit standards. Then we started to get down to the nitty-gritty of designing an actual experiment. It was devilishly simple. Everyone knows how to play tennis. I mean, not everyone knows how to *play* tennis, but everyone knows what's involved in playing tennis. You stand holding a racket and wave your arms around in the air trying to hit a ball. John McEnroe might not forgive me for that description, but that's pretty much the central thing in tennis: *waving your arms around in the air.* And that's all we needed—something that was easy to convey ("Imagine playing tennis"), but that would result in people imagining a similar but complex series of movements.

It worked like a charm. Melanie spent the next three weeks scanning twelve more willing participants who imagined playing tennis in the scanner and found the results were reliable and consistent. Every participant activated an area on the top of the brain known as the premotor cortex. *Every single subject.* All exactly the same.

We couldn't have hoped for a more reliable response if we'd asked our twelve healthy participants to all raise their right arm in the air. In fact, I have asked audiences to do this many times in my lectures, and because some people don't know their right from their left, the result is actually *less* consistent. Think about it—imagining playing a game of tennis more reliably produces a spot of activity in one part of your brain than if I asked you to raise your right arm. Why? Do we have a part of our brain dedicated to imagining playing tennis?

The answer is no, of course, but that this task works so well does have quite a lot to do with the game of tennis. We could have asked the participants to do anything that involved waving their arms in the air—holding two paddles and guiding a plane to park at a landing gate, for example. In principle that would have worked just as well, but I doubt that scenario is as universally familiar as tennis is.

What about another sport? Soccer is more popular than tennis and therefore more likely to be familiar to more people. The problem is there any many different ways to imagine playing soccer. Am I a striker, striding down the field and plunging the ball into the back of the net? Am I the goalie, diving left and right to quash the oncoming attack? Am I a fearless defender, sliding in for a tackle? All of these imagined actions will produce very different patterns of brain activity.

Tennis has one fundamental difference. Like soccer, there are many different aspects to playing tennis (serve, volley, smash!), but they *all* involve vigorously moving your arms. This common denominator is what made tennis imagery so perfect—its consistency and commonality. And imaginary tennis has one additional property that made it special—once you start, it's easy to keep doing it for thirty seconds, the time we needed to get a good scan. I remember asking one of our first volunteers how he found being asked to imagine playing tennis in the scanner. Quick as a flash he replied, "It was great—I won three sets to two!"

Of course you do need to know a little bit about tennis for this to work. If you have never heard of the game, then the instruction "Imagine playing tennis" will be meaningless and produce no discernible brain activity. But you don't have to be good at tennis for it to work. We have scanned non–tennis players, novices, and semiprofessionals, and almost without exception they activate their premotor cortex.

≈

We had what we needed. We had discovered that the two most reliable fMRI imagery tasks involved thinking about playing a game of tennis and imagining walking from room to room of your house. Imagining playing tennis was associated with robust fMRI activity in the premotor cortex; imagining walking around your house produced activity in an entirely different brain area—the parahippocampal gyrus.

To understand what happened next, it's important to know a little bit about where the premotor cortex is in your brain and what it does. Put your hand on top of your head. The premotor cortex is right there, a strip of brain in front of the motor cortex that sets up plans of action and comes into play whenever you initiate a movement. Think about what happens when you approach a door intending to open it and you reach to turn the knob. In this simple action, which you do more or less unconsciously, a cascade of motor programs are coordinated by your brain. As you approach the door, you reach out with your arm at just the right moment so your hand intersects the doorknob. You curl your hand into an appropriate shape to grab the doorknob (you'd do something entirely different if the door had a lever). Then you execute a simultaneous "twist and push" action with just the right amount of pressure to open the door—too little and the door won't open; too much and you run the risk of falling into the room and embarrassing yourself.

This is smoothly automatic, as are the thousands upon thousands of similar movements that are planned and guided each day by the premotor cortex. Because the premotor cortex helps set up these sequences, it is also activated whether or not we follow through with the sequence, or indeed if we only *imagine* the sequence. Set a coffee cup down on the table in front of you.

Think about what it *feels* like to be about to pick up that coffee cup. Now close your eyes and just imagine picking it up. You will find that it feels similar because planning an action feels similar to imagining that action, and the premotor cortex will activate in response to both.

≈

We were ready to test our new fMRI task on a patient like Kate. After years of preparation, the thrill of knowing that we could do it—at least in principle—combined with the uncertainty of knowing how long we'd have to wait for the right patient to come along, was spellbinding.

What happened next is the stuff of scientific fairy tales. Carol, a married twenty-three-year-old, was referred to us by her doctor from a town near Cambridge. In July of 2005, Carol had been hit by two cars while crossing a busy road. She suffered a traumatic brain injury and was admitted to a nearby hospital. A CT scan revealed brain swelling and substantial damage to her frontal lobes. Carol also had multiple lower-limb fractures. She required urgent care and underwent a bifrontal decompressive craniectomy. In this radical surgery, part of her skull was removed to allow her swelling brain to expand without being crushed by the inner walls of her cranium. The part of the skull that is removed is called the bone flap, and it is usually preserved because, if the patient recovers sufficiently and the brain swelling goes down, it can much later be replaced in a procedure known as cranioplasty. By September 2005, Carol's condition was considered stable. She was moved to a rehabilitation hospital closer to her family.

When I first met Carol, I was shocked by her condition. It's never easy meeting victims of a traumatic brain injury, but Carol's accident was still relatively recent and she looked awful. The decompressive craniectomy, lifesaving as it may be, is also visually arresting. Patients such as Carol look as if part of their head has sunken in; a shallow well of thin skin rests lightly on the surface of the brain. I have had to prepare many students for this sight before they met their first trauma victim, and I suspect many of them never fully recover from it. It was hard not to feel immensely sad for Carol. Whatever happened, even if she made a complete recovery, her life would never again be the same. In a single, deadly instant, two cars and a moment's distraction had redefined the rest of her life. She was a shocking reminder of how vulnerable we are and how quickly our lives can change.

Carol had lain in a hospital bed for months without responding or showing the slightest sign of inner awareness. Compared to the patients we were now seeing regularly, she was unremarkable. She had repeatedly been tested by experienced neurologists and diagnosed as vegetative. We didn't select her for any reason other than that she was the next in a line of patients who fulfilled all the requirements to go in the fMRI scanner.

We were starting to get some recognition for what we were doing—the publicity surrounding Kate's case had helped to generate interest from around the UK, and the scientific papers that we had published describing Kate, Debbie, and Kevin had attracted the attention of several other hospitals, which referred a regular flow of patients, sometimes one or two a month, who would travel to Cambridge by ambulance to be scanned by our team. But we were finally ready for something completely different. We were going to ask Carol to *do* something. This required that we give her instructions—tell her what we would like her to do and when. Until that point, we had simply done things *to* patients; showed them faces, played them words or whole sentences. All they had to do was just lie there and (we hoped) absorb what we were trying to convey. But we wanted Carol to follow a command, to activate her brain in certain ways in response to our instructions.

We asked Carol to imagine playing tennis; we asked her to think about swinging her arm back and forth, a volley here, a drop shot there, and perhaps an occasional smash. We repeated these instructions five times. We wanted her to imagine that she was playing tennis as if her life depended on it. As though she were playing match point, center court, in the finals at Wimbledon!

As the instructions were read to her one last time over the intercom, the atmosphere in the control room was tense. Did this make any sense? In one way it felt like total madness. *We were asking a vegetative-state patient to imagine that she was playing a game of tennis!* But inside the scanner, something amazing was happening. Whenever we asked Carol to imagine playing tennis, she would activate her premotor cortex just like healthy volunteers! When we asked her to stop—to just relax and "empty her mind"—the activity in the premotor cortex disappeared. Incredible, to say the least!

We then asked Carol to imagine walking around her home. Again, we asked her to do this five times. We wanted her to take herself back to that place where she had spent every day of her life before the accident. We wanted her to think about the layout of the house, to move from room to room, visualizing the furniture, pictures, doors, and walls.

We knew we were asking a lot, but Carol was obviously up for the task. When we told her to walk from room to room, her pattern of brain activity was identical to that of healthy volunteers. When we told her to let her mind go blank, she did that, right on cue. It reminded me of cheesy medical dramas when doctors ask patients, "Squeeze my hand if you can hear me." But we weren't asking Carol to squeeze our hand. We were asking her to activate her brain. And she was doing it! Kate's words echoed through my head: *Keep up the brain scanning. It was like magic, it found me*. This time, it really was like magic. We had found Carol. She wasn't vegetative at all. She was responding to us, doing everything we asked.

I was ecstatic. Carol was conscious and *we knew it*!

This thrilling "eureka" moment came after years of experimenting, refining, tweaking, and thinking, tunneling down and down, chipping away at the problem, hoping that the answer lay around every next corner. And we were there! We'd found the mother lode.

It may seem odd that we didn't then just go blazing ahead, scanning Carol day in, day out to find out what her world was like and, perhaps, improve her quality of life. Unfortunately, that's not how science works. Our only way to push the science forward was to stick to the strict protocols that we had established beforehand with our ethics committee—protocols that would be scrutinized and approved by the wider scientific community when Carol's story was eventually published in a scientific journal. With Carol, our stated goal had been to detect consciousness, not to haphazardly engage her in a tête-à-tête. We had invested an enormous amount of money and energy, scientific capital if you will, to get to this point and move the field forward. We were playing a long game—Carol and our other early patients were the first pioneers who would make contact with people in her situation possible, not to mention cast a new bright light on the nature of consciousness itself.

≈

It is, perhaps, ironic that Carol's family were never explicitly told that we had detected a conscious mind in her. We wanted to tell the family, but we simply weren't prepared. When we'd applied to the ethics committee to do this research, we hadn't even considered the possibility that we'd find a conscious person and, if so, what we'd do about it. Even small changes to the protocol, such as the number of scans that you intend to do on each patient, need to be cleared with the ethics committee in advance. Here we had much more than a change of protocol—it was a whole new reality! The principle at the heart of this rule—that every research study is scrutinized in advance by an impartial ethics committee—is good, as frustrating as it was for me at the time. Imagine, for example, that we had told Carol's mother that her daughter was conscious and locked inside her own body, and Carol's mother was so distressed by this news that she killed herself. Imagine it made Carol's husband so angry that he murdered the driver of one of the vehicles that had run Carol down five months earlier. These are dramatic and unlikely outcomes for sure, but were they to happen, who would be responsible? A more likely scenario was that the family's attitude to Carol would change, and the consequences of this also needed to be thought through carefully in advance. Would they understand that being conscious did not necessarily mean that her likelihood of recovery was greater? Would we be giving them false hope? Would they realize that, although we had made contact and established that Carol was conscious, right now that was all we could do? There was no cure, no solution, and no way to communicate with Carol on a regular basis. We hadn't thought through any of this because we didn't know that we were ever going to find an entirely nonresponsive patient who was conscious.

≈

In the end it wasn't my decision. I was just the guy who asked the scientific questions and then devised the methods to answer them. Our ethics protocol permitted the scans, but made no mention of what we would tell the family if we found a patient such as Carol. Carol's future care was a clinical matter, and I had no authority to interfere with that. If her family was going to be told, then it would have to come from the attending physician, who in this case decided that telling the family would not be clinically beneficial for Carol. I suppose he felt that the burden of knowing that Carol was in there, conscious and aware with no way of expressing herself, was worse than the burden of just not knowing, or assuming that Carol had no inner life at all. Or perhaps he felt that the ethical can of worms that cases such as Carol's opened up was not something worth addressing—less urgent perhaps than ensuring that her medical condition remained stable. I disagreed. I remembered Kate and Debbie, both of whom had experienced some improvement in their conditions after their families learned of their positive scans, and I couldn't help but wonder whether the same might be true for Carol and her family. But that wasn't enough to convince the attending physician. It was heartbreaking.

Nevertheless, Carol sparked my interest in the ethical complexities and legal issues in doing science with this unique group. I resolved to tackle some of the questions that her case had presented by engaging with philosophers and ethicists who understood the complexities of these issues. It was all I could do to make sure that this situation never occurred again. Carol was returned to her home town, and I never saw her again. There was no point—we had found her, but at the time we could do nothing further about it. She died in 2011 of long-term complications arising from her injury. Ironically, I was given this information by her attending physician.

≈

A single-page article describing our results appeared in *Science* in September 2006. A media storm erupted about the "vegetative patient who turned out to be aware and locked in her body." But Carol remained the anonymous hero. It provoked wonder and disbelief. We had made contact with a thinking person. A person who could imagine playing tennis and walking through her house. I was sure Carol could imagine and remember. I was sure she could still hope and dream.

On the day of publication, all three of the major TV stations in the UK turned up at the Unit for interviews, and we made the evening news on every channel. We were on the front page of every major UK newspaper and hundreds of foreign publications, including the *New York Times*. I was assigned a media person from the MRC's head office down in London, who fielded calls, choosing which I should respond to. It was bedlam and it went on for weeks and weeks. CNN's Anderson Cooper stopped off on his way back from an assignment in Africa to interview me for a *60 Minutes* special. He wanted to be scanned, so we scanned him. I asked him to imagine playing tennis in the scanner, and just like Carol's, his premotor cortex lit up on demand. For several months I didn't do much else but talk on the telephone or to cameras.

But there was something deeper than all the media attention that was ultimately more compelling and scientifically fulfilling. Something about the person that we'd found. Carol had been willing to reach out, even after what had happened to her, in her incomprehensible and broken state. Beyond the veil of her physical injuries, a sentient person wanted to make contact, to communicate, to say, "I am here," "I *exist*," "I am still *me*."

Carol was hopelessly disadvantaged by her useless body but was nevertheless still in there—her personality, attitudes, beliefs, moral compass, memories, hopes and fears, dreams and emotions. And perhaps most affecting of all, she had a will to respond, to reach out, to be heard. Carol had reached out to us. And we had found her.

≈

Over the next few months, e-mails flooded in from my peers, interested onlookers, and complete strangers. Broadly speaking, they all either said, "This is amazing!" or "How could you possibly say this woman is conscious?"

The skepticism confused and intrigued me. I knew that we had sent a clear "Are you there?" signal into inner space, and the answer "Yes, I am here" had come back loud and clear. I had no doubt that Carol was conscious—a thinking, feeling person trapped within a useless body. How could anyone dispute that? But they did.

The main objection was simple: Carol was in a vegetative state and entirely unaware of anything, yet somehow our instruction to "Imagine playing tennis" had fired off an automatic response in the premotor cortex that we had mistaken for a sign that she was conscious and willingly obeying our instructions.

It is still easy for me to see why some people preferred this explanation to our own—the idea that a patient who everyone thinks is vegetative is actually conscious and trapped inside her body is horrific. So horrific that for many of us it is entirely beyond our comprehension—our minds can't accept it as a possibility. Yet, that is the truth we had found, and like it or not, we had to fight for it. Suddenly, we knew what no one else knew, and I felt an intense responsibility to tell the world. Not all of these people are what they appear to be! At least some of them are thinking, feeling people!

The grim reality for the thousands of patients and their families, families like Maureen's and Kate's and Carol's, came into sharp focus for me right there and then. For years, many of these patients have been *warehoused*—an unfortunate term that is often used to describe their being permanently placed in environments without the expertise to carefully assess their mental functioning. And now we knew that some of these patients were likely to have been completely conscious all along. The thought still makes me immensely uncomfortable, as I suspect it does for many of you. I had to do something, not just for Maureen or any of the patients that we had scanned, but for the thousands of voiceless people who hadn't made it into a scanner to make themselves heard.

≈

Once the barrage of media attention around our successful attempt to communicate with Carol had finally started to subside, I focused on defending our scientific findings. The main problem with our detractors' arguments was lack of any evidence that their theory was physically possible. No one has ever shown that an unconscious brain can generate an automatic response on cue to a specific command. The brain does respond automatically all the time. When you hear the sound of a singing bird, your auditory cortex lights up whether you like it or not. A shining light on a dark night stimulates your visual cortex before you are even aware of it. The face of a friend in the crowd elicits an automatic flicker of recognition in your fusiform gyrus. Carol's response was something else. Our premotor cortex does not automatically light up when we hear the words "Imagine playing tennis." To be blunt, it only lights up *if we want it to light up.*

To prove this we carried out another experiment—probably the daftest I have ever conducted but entirely in keeping with Unit quirkiness. We put six healthy participants in the scanner and told them, "We are going to tell you to imagine things. Please just ignore what we ask you to do." Then, with the scanner running, we carried out exactly the same procedure that we had initiated with Carol. The participants heard "Imagine playing tennis" and we waited to see what would happen. There was no response. Not a flicker of activity from the premotor cortex in a single person!

Although these six people had been explicitly told to imagine playing tennis—exactly what we had told Carol to do—they didn't do it because they had previously been told not to do whatever we were about to ask.

This was rock-solid evidence that being asked to "Imagine playing tennis" was not enough to fire off an automatic brain response, let alone activity exactly where we predicted it would be, in the premotor cortex. Carol's brain had responded as it did because she had *wanted* it to. She had responded because she was conscious.

I was proud of our daft little experiment, although the arguments against our conclusions held no water for many other reasons. First, what was most remarkable about Carol's response was that she was able to sustain it for the thirty seconds that we needed to get a good scan. In spite of receiving no additional instructions or encouragement when she heard the words "Imagine playing tennis," Carol activated her premotor cortex and kept it activated for a full thirty seconds. Of all the "automatic" brain responses we know about (to sights and sounds, for example), none is sustained in the absence of additional stimulation. When you hear a single gunshot, your auditory cortex responds immediately. But thirty seconds later that response will be long gone. But because Carol's responses reflected her own mental imagery and because we know that people can "play tennis in their heads" for thirty seconds or more without interruption, Carol was able to generate a sustained response that could have only occurred if she was conscious.

The final argument against those who doubted our interpretation of Carol's brain activity was a more philosophical one. After a severe brain injury, when the request to move a hand or a finger is followed by an appropriate motor response, it is taken as a sign of awareness. By analogy, if the request to activate the premotor cortex by imagining moving the hand is followed by an appropriate brain response, shouldn't we give that response the same weight?

Skeptics may argue that brain responses are somehow less physical, reliable, or immediate than motor responses. But, as is the case with motor responses, these arguments can be dispelled with careful measurement, replication, and objective verification. For example, if a person who was assumed to be unaware raised his or her hand in response to a command on only one occasion, some doubt would remain about the presence of awareness. The movement might have been a chance occurrence, coincident with the instruction. However, if that same person repeated this response to the command on ten separate occasions, little doubt would remain that the patient was aware. By the same token, if that person was able to activate his or her premotor cortex in response to a command (by being told to imagine playing tennis) and was able to do this in every one of ten trials, would we not have to accept that he or she was aware?

Fortunately for us, Carol's brain activity had not been a one-off. She had activated her premotor cortex when asked to imagine playing tennis and her parahippocampal gyrus when asked to imagine moving around her house on multiple occasions during the scanning. The case was closed. Carol was conscious.

≈

Carol turned the whole notion of the gray-zone vegetative state on its head and presented a new and significant challenge for physicians all over the world. MDs everywhere started to think again about patients in their care. Had they made the right diagnosis? Was there a chance that one of their patients was still in there, like Carol, despite all appearances to the contrary? Inquiries came from the most unlikely places. What did it mean for medical insurance? How *would* you insure against that? What about legal decisions regarding life-sustaining therapy? If Anthony Bland, the Brit injured in the soccer stadium stampede, had been able to imagine playing tennis, would he be alive today? What about Terri Schiavo?

Carol had made it undeniably clear that some patients who appear to be vegetative may be entirely aware of the world around them and able to generate sequences of responses when asked to do so. Was this another gray-zone state? Perhaps or perhaps not. Do these people spend periods of their trapped lives completely unaware and other periods entirely conscious and cognizant of everything going on around them? We didn't know, but we were beginning to home in on the building blocks of cognition, a kind of critical mass of flickering, tenuous neural connections that in some patients seemed to be firing sporadically, trying to reignite, perhaps indefatigably forging new pathways in a moribund brain.

≈

I had kept in touch with Phil, Maureen's brother, and we had gone to several more gigs together over the years. Each time we met, he reported that Maureen's condition was unchanged. His parents, Isa and Philip, were trying to take each day as it came.

In 2007, Phil and I went to see the Waterboys at the Corn Exchange in Cambridge. It was particularly bittersweet. The album that had brought the band their first big surge of recognition, *Fisherman's Blues,* had been released the year Maureen and I fell in love. It had been the sound track to our overwhelming passion and our struggles.

Around this time, Maureen's father, Philip, wrote to me. He explained that Maureen's doctor had agreed to put her on an experimental trial of the sedative zolpidem (also known as Ambien), primarily used for the treatment of insomnia. In 2000, a case report in the *South African Medical Journal* described a young man who "awakened" within thirty minutes of receiving zolpidem after three years in a vegetative state. Philip had tried the drug on Maureen, and her doctor had been convinced that she had responded positively: "Her facial expressions are less strained now and she appears more aware," he reported.

Philip was less optimistic: "I have been unable to convince him [Maureen's doctor] that the hand movements he observed and the squeezing of the hand/fingers are things that Maureen does without any request being made."

I remembered that Maureen's father was a scientist, and I implicitly trusted his judgment. Maureen's doctor spent but a short period each week with her, while Philip had much more opportunity to collect reliable data by observing her daily.

I asked Philip to send me video recordings of Maureen on and off zolpidem. Two VHS tapes soon arrived in my mailbox. This was science—not science in the laboratory but science in the real world. I slipped the first tape into the VCR. There was Maureen, the same woman I had known and loved. All the efforts of her parents, which Phil had told me about, the daily massages and impeccable grooming, were in evidence. There was no spasticity, no alteration in her appearance. She looked remarkably intact and unchanged, her wild chestnut hair, shorter than I remembered it, resting lightly on the pillow, the lovely face so given to laughter and strong opinions smooth and unconcerned.

I watched the two tapes carefully from beginning to end, then I watched them again. I switched them up and tried to tell them apart. I couldn't. As desperate as I was to see an improvement on the drug, there was none. At least not when I did a carefully controlled "blinded" study in the comfort of my living room.

I e-mailed both Philip and Maureen's doctor: "I did take a good long look at the videos and also went through your detailed account of the findings with Maureen. The results are not at all encouraging. My various correspondences with other clinicians who have tried zolpidem in different patients are overwhelmingly disappointing. The responses observed are, for the most part, very minor, transient and in some cases are difficult to disentangle from the likely effects of the increase in encouragement and stimulation from the family that these trials typically engender."

Almost ten years on, it appears that my English reserve was probably entirely appropriate. The South African case resulted in countless trials of zolpidem, and few resulted in consistent results in vegetative patients. A comprehensive recent study by my friend and colleague Steven Laureys in Liège, Belgium, failed to show an improvement in even one of sixty patients with disorders of consciousness who were tested on the drug.

When I next met Phil, he said of my BBC appearance in the wake of the publication of our tennis experiment and Carol's results, "That must have been nerve-racking!"

I told him that I was getting used to media attention, which I felt was important to raise awareness about people like Maureen. He thanked me and we moved on. But I kept replaying that quick exchange in my head. By exploring the gray zone, was I trying to make things right with Maureen? Did I need to get to a place of forgiveness and understanding? Had something unresolved in our embattled relationship been driving me all along?

CHAPTER NINE

YES AND NO

≈

As all the Heavens were a Bell,
And Being, but an Ear,
And I, and Silence, some strange Race,
Wrecked, solitary, here—

—Emily Dickinson

We tried our tennis technique on as many patients as possible to see if it worked reliably and to improve it. By 2010, in collaboration with Laureys, we had scanned fifty-four patients performing the tennis and spatial-navigation tasks. Given the thousands of research dollars involved and the weeks and months of recruitment, assessment, replication, and verification, fifty-four successful scans was an incredible achievement by any measure. Of these, twenty-three patients had repeatedly been diagnosed as vegetative through intensive neurological examination. Nevertheless, in the fMRI scanner we found that four of these twenty-three (17 percent) generated convincing responses.

The long journey that had begun with Kate, more than ten years earlier, had culminated in a kind of vindication. As I'd long suspected, some of these patients *were* conscious. And not just conscious in the vague, foggy half-here-and-half-there sort of way that we all experience as we drift off to sleep at night, but conscious enough to listen to a set of instructions and turn those instructions into a deliberate, rather elaborate imagined activity for a full thirty seconds, which in turn generated a set of brain responses that we could detect with our powerful new generation of fMRI scanners. They were in there—just like you and me—watching, listening, awake, *and* aware. Yet, somehow, unlike you and me, they were stuck, trapped in the gray zone, lost in inner space, unable to break out unless they were one of the few lucky ones who made it into our scanner.

I began to think about those who weren't as lucky. How many were there? The implications were chilling. We don't know exactly how many vegetative-state patients there are. This is due, in large part, to poor nursing-home records. Estimates range between fifteen thousand and forty thousand in the United States. Our findings suggested that as many as seven thousand might actually be aware of everything going on around them.

We had a vocal contingent that disputed our findings. They argued that although 17 percent of our vegetative-state patients had been responsive in the scanner, only one of our thirty-one minimally conscious patients (3 percent) had produced the same kind of responses. Patients who appear to be minimally conscious are generally less severely brain damaged than patients who appear to be vegetative. Why then would they be *less* likely to show responses in the scanner? It didn't make sense. Surely they should be *more* responsive.

Six years on we would know the answer to this question, but at the time it was puzzling. As it turns out, most minimally conscious patients are more or less what they appear to be—*minimally conscious*. It's often unclear exactly what that term means—it's hard enough to get scientists to agree on what consciousness is, let alone define what *minimal consciousness* means. But let's just say that being minimally conscious means that sometimes you're there, sometimes not, and sometimes you're stuck somewhere in between. Either way, at best you can give some subtle signal—perhaps the movement of a finger—to say you're there. At worst, you can't even do *that*. It's not surprising that few of our minimally conscious patients were able to follow a set of instructions in the fMRI scanner and turn them into the complex sequence of mental acrobatics that are required to imagine playing a game of tennis. Why should they be able to? Most of the time, they weren't even able to move a finger reliably, so why would they be able to imagine playing tennis? For the nineteen vegetative patients who also couldn't imagine playing tennis, the situation is similar, only worse. They lie unawake, unaware—in a part of the gray zone so remote and murky that even they don't know they're there. Of course they can't imagine playing tennis—they can't even think!

But what about the fantastic four? What about those four patients who appeared to be vegetative, yet *could* perform these amazing mental feats in the scanner? They were something quite different, something quite special. In fact, they were not *vegetative* at all. They were not even *minimally conscious*. They were in a state—a part of the gray zone—for which we still have no name. And in that part of the gray zone, you can be completely awake, completely aware, yet completely physically nonresponsive—unable to blink an eye, raise an eyebrow, or move a muscle. It was not at all surprising to me that these four patients could imagine playing tennis. No more surprising than that you and I can too.

Our findings had raised an even more interesting possibility that was already starting to excite me. A lot. State-of-the-art changes in computing technology had produced scanners that were now capable of revealing a life lived inside an unresponsive body, and the possibility of a real brain-computer interface was starting to emerge—a machine that was capable of providing a bridge between the gray zone and the outside world. Asking patients to respond by imagining playing tennis was one thing, but could we use these incredible new tools to actually communicate with them?

Working with Martin Monti—one of my bright, confident postdocs at the Unit—we devised a way to make two-way communication possible. As usual, we started with a series of whacky experiments in healthy volunteers—in this case, me. Martin is Italian and Jewish, raised in Italy and educated partially in the United States. This unusual combination was particularly useful a couple of years later when I was asked to consult on the politically charged case of Israeli prime minister Ariel Sharon, who suffered a stroke in 2006 and spent eight years on life support until his death in 2014.

While Sharon was incapacitated, one of his people got in touch with me through an Israeli colleague and asked me to come to Israel and scan him to see if beneath his unresponsive exterior he had preserved awareness. I was happy to help. But try as I might, I couldn't get another member of my team to accompany me.

"Why is Sharon any more deserving of our time and attention than the patients we have much closer to home?" they argued. I could see their point. All Sharon had over the patients we were seeing day in, day out was that he was famous and the ex–prime minister of Israel. Did that somehow make his life any worthier or his condition any more important than theirs? Traveling to Israel would draw significantly on our time and resources, and it wasn't at all clear that those resources wouldn't be better spent on our own, more local patients. But I suspected there was more to it than that.

"Assessing a famous patient could raise the profile of the lab and would bring attention to this population of patients and their plight," I said. A significant part of my life was being spent talking publicly about patients with disorders of consciousness, and I was keen to teach my students and postdocs about the benefits of maintaining a good relationship with the media.

"Not if he's a war criminal," came the reply.

I googled Ariel Sharon. Sure enough, reams of pages argued that point. Reams of pages also argued the opposite, but I wasn't about to let political opinions divide my lab.

I contacted Martin, who had taken a job as an assistant professor in the Department of Psychology at UCLA, and in 2012 he traveled to Israel and scanned Sharon. He reported back to me that Sharon's scans showed fairly basic responses—nothing high level. He had asked Sharon to imagine playing tennis and also to imagine he was walking through the rooms of his home. As Martin said to the press at the time, "Information from the external world is being transferred to the appropriate parts of Mr. Sharon's brain. However, the evidence does not as clearly indicate whether Mr. Sharon is consciously perceiving this information."

In truth, the results were inconclusive. As Martin put it, "He may be minimally conscious, but the results were weak and should be interpreted with caution." Sharon was like many of the patients that we have seen over the years—with some evidence for a response but no clear evidence for consciousness. Just like Kevin, Debbie, or Kate. But there was a difference. When we'd scanned Kevin, Debbie, and Kate, we hadn't known how to reliably detect consciousness even if it was there. We were left trying to decide whether the rather basic responses that we saw to words, sentences, and faces could possibly reflect covert consciousness. In Sharon's case, Martin had given him the acid test—the test that we now knew could detect residual consciousness in an entirely unresponsive body. And the test had returned a negative. Sharon had not been able to imagine playing tennis—at least not in a way that left Martin able to draw definitive conclusions. "The results . . . should be interpreted with caution." I've lost count of the number of times I have had to say that to an attending physician or a distraught family member.

Sharon's case raised many difficult questions. For example, during the period he was incapacitated, he had surgery to treat a kidney infection. People objected to what they felt was excessive care for someone with a severe disorder of consciousness.

Judaism takes the position that all human life is sacred and is to be protected at virtually any cost. As Rabbi Jack Abramowitz wrote in an interesting blog on the subject in 2014, "If a person will die from fasting on Yom Kippur, not only may he eat, he is required to do so. Similarly, in a life-threatening situation, one must violate Shabbos to call an ambulance or take someone to the hospital."

An interesting corollary of this is that Judaism has no concept of "quality of life." A healthy person is not more entitled to kidney surgery than a minimally conscious patient. It's an interesting perspective, but not one that I feel great affinity with. Some decisions are certainly tougher than others. For example, it's hard to decide whether a teenager with cancer is more deserving of treatment (in a situation where you have to choose one or the other) than a young businessman with a major head injury whose company is pioneering a new energy-saving lightbulb. These sorts of arguments have kept many a philosophy graduate student awake at night. But at the extremes, it feels much simpler to me. A teenager with cancer versus an eighty-five-year-old minimally conscious patient with failing kidneys? That wouldn't be a difficult decision for me. The world doesn't really work like that—when one person is given treatment, it's not generally the case that someone, somewhere else, is denied it. But at some level it must be true. The decisions we make today have consequences for others, far away in space and time. Consequences that most of us don't even realize.

All of us are different and our personal circumstances play an important role. If forced to choose, Ariel Sharon's family may—understandably—value his life over that of an anonymous teenager with cancer. What role then, should society or religion play in dictating how we make such decisions when no one size fits all? Can we do better than utilitarianism? Is it possible to gauge the absolute social good in such a situation? Should social factors come into play at all? Perhaps this is why Judaism discounts utilitarianism all together, saying those type of judgments and decisions belong outside the human realm. Still, human beings make them, so I'm not at all sure, in a practical sense, how useful that position is.

≈

Back at the Unit in 2010, long before Ariel Sharon went into the scanner, Martin and I were working day and night to devise a simple method for communicating with fMRI. I'd been convinced for some time that two-way communication could work with fMRI, but eventually I decided to put it to the test myself. Some scientific questions are so fundamental, so basic, that it's just easier to ask them of yourself rather than wait around for an experiment involving tens of participants, hours and hours of scanning, and lots of paperwork. It just isn't worth it. In this case, all I cared about was whether it was possible for me to communicate with the outside world by changing my pattern of brain activity inside an fMRI scanner. I handed Martin a piece of paper with a series of questions scribbled on it—questions that he could not possibly know the answer to. He knew me, but not well enough to know the answers to such questions as "Is my mother still alive?" Or "Is my father's name Terry?" The questions were unimportant; they just had to be obscure enough that Martin wouldn't already know the answers, yet simple enough that they could be answered by me with a simple yes or no.

I lay back and closed my eyes, listening to the whirring of the scanner bed as it slowly pulled me into the fMRI. It was warm and dark inside. The bore—the long tube through the center that ran the length of my body—was less than two feet wide. My elbows just about touched its sides. A wool blanket lay over my legs, and my head was immobilized by small sponge cushions that the technician had jammed down between my skull and the head coil. The "coil" is a bit like having your head inside a birdcage. You can see but only by peering through the spaces between the "bars," which are positioned right in front of your face. When you climb into the scanner, the birdcage lies open like a clamshell. You lie down, placing your head into one half of the birdcage, and the technician brings the other half down over your face, locking your entire head inside. These birdcages are the receivers and transmitters of the radio-frequency signals that are the essence of MRI technology. They are constructed to lie close to the head because that vastly improves the image quality.

I knew I had ten minutes or so while the technician went through the necessary setup procedures. As I lay there in the dark I started to think. I'd been inside a scanner before, many times. In fact, I'd been inside many scanners long before I ever knew that they would become such a fundamental part of my life. When I was fourteen years old, I'd been diagnosed with Hodgkin's disease. I spent the best part of two years in and out of scanners, MRI, CT, ultrasound, X-ray—I'd had them all. In 1981, I spent a few minutes each day for seven weeks inside a linear accelerator, a huge machine that filled an entire room and delivered bursts of radiotherapy to my chest. Back then, these machines terrified me, despite the role they undoubtedly played in my treatment and eventual recovery. Choosing a career that involved spending so much time in and around scanners was an odd choice, I suppose.

Hodgkin's disease is very curable now, but back then it was a different matter. I don't know if I ever thought I would die, but I do remember that on many occasions I felt as if I were dying. As well as radiotherapy, I had many courses of chemotherapy. I went into remission. But the disease came back again, and I returned to the daily treadmill of injections, pills, and vomiting. I thought it would never end. My hair fell out, I lost almost half my body weight, and at times I just wanted to curl up and die. Some of my close friends did. Eventually, my duodenum—the first part of the small intestine immediately beyond the stomach—had enough of the drugs and gave up completely. The pain was unbearable. I was put on pethidine, an opioid from the same family as heroin and morphine.

Every four hours, I would collapse into unconscious ecstasy as the drug filled my veins, moving up my arm in a warm, comforting wave of relief. Then, like clockwork, three hours later I would wake, sit bolt upright, and endure another hour of excruciating pain until it was time again for my next bout of sweet relief. Eventually I began hallucinating—I danced through fields with dwarfs and pixies and held birds in my hand as they sang sweet songs. I was taken off pethidine immediately and came back to earth through a horrible, sweaty haze of pain and confusion.

During that period, I often felt that I was at the borderlands between life and death, my own kind of gray zone between being not quite here and not quite there. I came and I went, in and out, back and forth. I wanted to be there, not here, because in the gray zone I could escape the pain and sleep through the confusion. Each time I circled back to earth, out of the gray zone and back into reality, I'd scream obscenities until a kindly nurse would come to my rescue and deliver me back to the comfort of that place.

Despite the horror, throughout that period I was surrounded by a kind of spirit and love that has stayed with me ever since. My mother was there at my bedside every day for two years, cheerfully reading me the newspaper, updating me on the latest family gossip, and generally keeping my boat afloat. My dad came to the hospital every morning to deliver the newspaper, every lunchtime to share a cake or a joke, and every evening to wish me good night before taking the late train home. My brother and sister just had to get on with their teenage years, doing the best they could—I have no idea how they got through the awfulness of it all.

It didn't occur to me until many years later how unbearable this must have been for all of them. It was always about me. I was the patient, I was the one who was suffering, and I was the one whose future was uncertain. But in reality, it never is like that. Life-threatening illnesses affect us all. Their reach is virtually infinite. Like in a butterfly effect, when one member of a close family goes down, the ripples of turmoil escalate outward in a multitude of different and unpredictable ways. Close families often fall apart, regardless of whether the patient at the center of it all lives or dies. Fortunately, mine didn't, and I am still here to tell the tale.

Almost forty years on, I look at the faces of the mothers, fathers, brothers, sisters, and children of people in the gray zone and I feel some kind of affinity with them all; a sense of knowing what it's like for a family when the life of someone you love is on the line.

Lying there in the scanner thinking back on my childhood illness, I started to wonder about the choices I had made in my life and the possibility that it was all somehow inevitable that I would end up doing this. I'm an atheist and I don't believe in fate. But I do believe that the path we take is dictated by the choices we make, and those choices are informed by our experiences. I had been very ill as a child, and I was cured by the machinery of modern medicine. By drugs, by scanners, and by people who worked hard to keep me alive. Scientists, doctors, nurses, hospital porters—hundreds and hundreds of people who worked directly, and indirectly, to keep me going in the face of grave uncertainty. Now here I was on the other side. Was I trying to give something back? I had chosen to work at the frontiers of modern medicine, beside engineers who were developing the next generation of brain scanners, neuroscientists who were cracking the code of complex neurodegenerative diseases, and neurointensive specialists who were working day and night to bring both young and old back from the brink of death. Could that all have come about by chance? And what about Maureen's accident? Surely that's what had first piqued my interest in the vegetative state and conditions like it? And Kate? Had she not responded, I wouldn't be here, right now, in this scanner, trying to communicate with Martin. Perhaps it was inevitable after all that I would end up here.

≈

"Okay, we're ready. What now?" Martin's voice crackled in my headphones over the rudimentary intercom system that was my only means of communication with the world outside.

"Ask me one of the questions. If the answer is yes, I'm going to imagine playing tennis, and if the answer is no, I'm going to imagine walking around my house."

Ten seconds later I felt the scanner click, bang, and beep into action. It's complex physics, but it relies on spinning protons in the brain. As I was rolled into the bore of the scanner, the enormously powerful magnets above and around my head brought all of the protons in my brain into alignment (thankfully, I was blissfully unaware of any of this at the time). Then the birdcage around my head released a short burst of radio waves, knocking all of these protons out of alignment. After the radio-wave burst ended, the huge magnets pulled all of the protons back into line. The rate at which the protons in blood realign after being knocked over on their side depends on the oxygenation level of the blood, and this produces a signal that can be picked up by the scanner. Incredible technology, incredible science.

Being inside an MRI scanner is a curious thing. It's incredibly loud—so loud that you would suffer hearing damage if you did not wear in-ear plugs as well as the sort of earmuffs that you see on the heads of guys drilling on roads. There I was, lying inside a $6 million cocoon, pondering my childhood illness, with my head trapped in a birdcage and a noise that was as loud as a jet plane flying right past my ear. In that context, hearing Martin ask "Is your mother still alive?" was almost surreal. I had to think fast. I knew what I had to do, but I only had thirty seconds to do it. The answer was no, my mother was no longer alive, and to convey a no I knew I had to think about walking through my home.

I quickly turned my thoughts to coming in through my front door into the entryway of my small house near the center of Cambridge. I visualized the entryway, chockablock with coats and shoes. I walked on into the dining room. There was the glass table that I had bought from IKEA a year earlier. I noticed the matching and maddeningly uncomfortable chairs. I looked toward the kitchen, with its crooked hundred-year-old doorway. I went inside, past the fridge on my right and the door to the patio on my left. Straight ahead of me, I could see right through the back window into the garden. To get there, I'd have to turn left, go through the back door, across the stone patio I'd laid earlier that year, and onto the grass. In my mind, that's where I was headed.

"Now just relax and clear your mind."

Those words interrupted my train of thought, stopping me dead in my tracks. I quickly turned my attention away from my house. I'd asked participants to "relax and clear your mind" a thousand times before, and in that instant I realized what a ludicrous request that was—what does *clear you mind* mean? How can any of us "clear our minds"? When I relax, my mind fills with plans for tomorrow, the shopping I have to do, and the meetings I have to attend.

I'm reminded of the numerous times I've been asked, "Is it

true we only use ten percent of our brains?" I have no idea where that ridiculous idea came from, but it's nonsense. Nevertheless, enough people have heard it that I (and I suspect every other neuroscientist on the planet) get asked it all the time. But if you looked at a PET scan, a particular type of PET scan known as a fluorodeoxyglucose (FDG) scan, which measures the baseline activity of the brain when it is just resting, you'd see that all of it is active, all the time. Some of it becomes more active when you think or do certain things (and that is the basis of an O-15 PET scan or fMRI), but when you just "relax and clear your mind," all of your brain is still active.

There is no sense in which we only use 10 percent of our brains, just as there is no sense in which my mind "clears" when I relax. But that's exactly what I had to try to do lying there in the scanner listening to Martin's voice.

I took my mind back to Sydney and imagined lying on Bondi Beach with my eyes closed. I imagined the warmth of the sun on my face and tried to stay focused—focused on nothing. Try thinking about nothing for a few seconds and you'll find out just how difficult it is. Your mind is like a hummingbird, constantly flitting from one idea to another, and it's all but impossible to put the brakes on and go blank. I've often thought that this may be why it's so difficult for any of us to imagine what it's like to be in a vegetative state. What's it like to think about nothing? We can't know because we've never experienced it. And we can't ever experience it. Not this side of the gray zone anyway.

"Is your mother still alive?" Martin's voice brought me back from Bondi. It was a relief to hear the words again—I could go back to my house in Cambridge, back where I had left off thirty seconds earlier, standing in the kitchen contemplating how I was going to get into the garden. It's a strange paradox that it's so much easier to imagine something than to imagine nothing. In the real world, doing something requires so much more effort than doing nothing. But in your mind it's the reverse. We are always switched on, monitoring the world around us, looking for things that we should attend to, scanning for environments to avoid. That's the default state of affairs. Turning it off requires effort.

We repeated the procedure five times, switching back and forth between answering the question about my mother and relaxing on the beach. It took exactly five minutes, then the scan was over. The sudden silence was a relief. But I was on tenterhooks. Had it worked? Had I been able to communicate with the outside world using only my brain? I couldn't wait until I got out of the scanner.

"Do you know the answer?" I blurted out, hoping someone was listening. I was desperate to know. But I was trapped, still inside the birdcage, completely detached from what was going on in the control room. Nothing. The tension was killing me.

"Did it work?" I yelled.

More silence. Then, the intercom crackled. "Your mother is no longer alive."

I couldn't believe it. "Are you sure?"

"One hundred percent sure! It's as clear as day. Your parahippocampal gyrus lit up like a Christmas tree, which means you were imagining walking around your house—that means you were telling us no, right? Your mother is no longer alive."

Until that point, I could never have imagined a scenario in which hearing the words "Your mother is no longer alive" would make me happy. Now I was ecstatic.

"Let's do it again!" I yelled. "Ask me another question!"

≈

By the end of the session I'd been asked three questions, and I'd answered all of them successfully using just my brain. When I was asked "Is your father's name Chris?," I again imagined walking from room to room in my house because the answer was no—my father's name is not Chris. Chris is the name of my older brother. But when I was asked "Is your father's name Terry?," I did something quite different. I imagined playing tennis, whacking the ball over the net toward my imaginary opponent. I knew I had to do this to convey a yes. My father's name *is* Terry, and by imagining playing a game of tennis I conveyed that to Martin, outside in the control room. I had told him my father's name *just by changing the pattern of activity in my brain.*

By this feat of technological wizardry, Martin had been able to read my thoughts. Not in the telepathic sense, at least not literally. But what I was thinking had been recoded into a pattern of brain activity that had been picked up by the fMRI scanner and represented on the computer screen in the form of brightly colored blobs that Martin was able to "read." *He had read my mind.*

The experiment had worked! We'd shown that we could use fMRI for two-way communication with someone locked down inside the scanner. We could ask questions and we could decode the answers simply by looking at what happened inside the person's brain. It was delightfully simple, but it gave us exactly what we needed.

≈

We needed to answer many questions before we could try it in patients. How reliable and robust was the technique? Could everyone do it or was I special in some way? I'd spent a lot of time in fMRI scanners and knew lots about how to best activate brains—perhaps that gave me an advantage, an edge over the person in the street?

To test whether I was a special case, Martin scanned sixteen strangers, using the technique we had developed: playing tennis for yes; moving around your house for no. Sixteen people, three questions each. It took a couple of weeks to complete the experiment. When we were done, Martin bounded into my office, beaming from ear to ear. I knew what the result was, it was written all over his face. Amazingly, just by looking at the patterns of activation in the brain in response to each question, Martin was able to correctly decode the answers to every single one of the forty-eight questions posed in the experiment. It worked! Reliable two-way communication with fMRI was possible!

Sure, each answer took five minutes of scanning to decode with 100 percent accuracy, but imagine if that was your only way to communicate? Wouldn't it change your life? Imagine if you couldn't talk, blink, or make yourself known in any way for years on end, and then this possibility came along—a technologically supercharged version of the old parlor game Twenty Questions that connected a thinking brain, silenced by physical disability, to the outside world.

≈

We soon had the chance to put this technique to the test. As part of our collaboration with Steven Laureys and his colleagues in Belgium, we learned about a twenty-two-year-old Eastern European patient, let's call him John (I never knew his name), who had been riding his motorcycle five years previously when he was hit by a car. A massive blow to the back of his brain resulted in widespread cerebral contusion—bruising of the brain that often leads to multiple tiny hemorrhages as small blood vessels spill their contents into the surrounding brain tissue. Steven's group carefully assessed John for a week, and he was repeatedly diagnosed as vegetative. Melanie Boly, who was back in Liège working as a resident in clinical neurology, put John into the fMRI scanner and asked him to imagine playing tennis. Despite his five years of nonresponsivity, in the scanner John showed clear signs of awareness—he was able to imagine playing tennis when asked to do so.

Steven called from Belgium. Should his team scan John using our communication technique? It took me no time at all to agree. This was the opportunity we had been waiting for. The following evening Melanie and one of Steven's students, Audrey Vanhaudenhuyse, would scan John and try to communicate with him using our new technique. Fueled with excitement, Martin jumped on the first train to Liège—he badly wanted to be there and I wanted him there too. By then he'd built up a lot of experience communicating with healthy participants in the scanner, and he'd been writing some smart computer code for getting the results out quickly and efficiently.

The day of the scan, I woke up, jumped out of bed, and reached for my suit and tie. I had a speaking engagement at a meeting of the Royal Society of London. I hadn't prepared at all—my mind had been completely preoccupied with the events in Belgium. As I sat on the train, slowly trundling toward London, I tried to focus on the talk that I had to give, but I kept thinking about John and his scan. I wished I could be there. Perhaps I should have gone? Although I had agreed to the talk in London months earlier and pulling out would have been entirely inappropriate, I can't pretend that I hadn't been tempted.

I was barely inside the Royal Society before my cell phone rang. It was Martin calling from the scanner room in Liège.

"He's responding," Martin yelped. "He's imagining playing tennis again. Shall we ask him a question?"

"Do it!" I yelled back over the din of the foyer crowds.

As I waited to give my talk, my cell phone rang every few minutes. "It looks like he's activating his premotor cortex, but we can't be sure," Martin informed me.

The Belgian scanner was identical to the one we had in Cambridge and could analyze fMRI data on the fly, but only in a superficial way—sometimes it was hard to be absolutely sure what the final result of the scan was.

"Can you get a better look at the raw data?" I asked. If Martin could get hold of the data and run his own analysis, I was sure we'd have a better idea what was going on.

I had to turn off my phone to give my talk, titled "When Thoughts Become Actions: Using fMRI to Detect Awareness." Forty-five minutes plus questions on my work on detecting awareness in the vegetative state. It was a tough audience—two hundred people attended, including many of the smartest cognitive neuroscientists in the UK, but the talk was well received and the audience seemed convinced. The moment I came down from the podium, I was back out in the lobby and on the phone to Liège. People tried to engage me in further questions about my talk—I shooed them away. My head was in Belgium, and I was on a knife's edge.

"They want to know what we should ask him," Martin said.

"Tell them to use the same questions that you asked the healthy participants. Ask him if he has any brothers or sisters."

"We've done that. We've asked all three questions already. What next?"

Things were happening so fast that we'd run out of questions. We hadn't even considered what to do if the patient got this far. I suppose we just didn't believe it would happen.

"Audrey wants to know if we should ask him if he likes pizza," Martin said. It was turning into a game of telephone, and I was becoming concerned that important details were getting lost in translation.

Audrey's suggestion raised an important issue. So far, we had only asked questions that had definitive yes or no answers that could be verified by interviewing the family after the scan. Questions such as "Have you got any brothers?" are definitive. You either have or you haven't. They can also be verified with family members. But questions such as "Do you like pizza?" are not. I like mushroom pizza, but I don't like pepperoni. My answer to the question, then, is "It depends on what kind of pizza."

In addition, my preference for pizza is not something that is a checkable, unassailable fact, such as whether I have a brother. We agreed that asking John his father's name was a good option, as well as where he had last gone on vacation before his accident five years earlier. The family was contacted and gave some possible answers, some right and some wrong, and Audrey went back to the scanner.

So it went. Steven's team in Liège scanning the patient, and me in London providing advice—for the first time in history we scanned and communicated with a patient who had been declared clinically vegetative. When the formal analysis came back from Martin, it was crystal clear that John had answered five questions correctly. Incredibly, he had indicated that, yes, he had brothers; no, he didn't have sisters; yes, his father's name was Alexander; and, no, it was not Thomas. He also confirmed the last place that he had visited on vacation before his injury—the United States.

We had time for just one more question. Perhaps it was time to push things further, to ask a question that we couldn't possibly verify, a question that could actually make a difference to John's life. Standing in the scanner control room, Martin, Audrey, and Melanie came up with an idea—they would ask John whether he was in any pain. If John had been in pain for the last five years, here was a chance to find out and perhaps even do something about it. Melanie called Steven on the phone for his advice. Steven was the local ethics specialist, and by then he was experienced at making decisions about what to do—and what not to do—in situations like this.

"Ask him if he wants to die," Steven said.

Melanie was taken aback. "Are you sure? Shouldn't we ask him if he's in any pain?"

"No!" Steven responded. "Ask him if he wants to die."

It was a harrowing moment. We'd decided to push things further than we'd ever pushed them before, and now we were facing the possibility of pushing them in a new—and frankly terrifying—direction. What if he replied yes? What would we do? Even if he replied no, we could do nothing much but accept that at least we now knew what his wishes were.

None of us, including Steven, had thought through the ethical conundrum that this situation posed. For almost ten years I'd been working toward this—working toward communicating with patients in the gray zone and asking them their wishes—but now that we were there, I had no idea what we were going to do with the answer. I wasn't even sure that we should be asking the question! But in Liège, Steven ran the show and the decision was his. I suspect he knew that ultimately this was the important question—the question that the family wanted to ask.

It's hard to say whether what happened next was good or bad—in many ways it got us out of a difficult situation, but I can't pretend that I wasn't disappointed. The results of John's scan when he was asked "Do you want to die?" were inconclusive. Despite answering the previous five questions clearly and accurately, John's brain activity when he was asked whether he wanted to die was impossible to decode. It wasn't that there was no response, it was just impossible to say whether he was imagining playing tennis or walking through the rooms of his home. He appeared to be doing neither. It was impossible to know whether his answer was "Yes, I want to die" or "No, I do not want to die." I have no idea why this happened, but I suspect that like "Do you like pizza?," for most of us "Do you want to die?" does not have a clear yes or no answer. Perhaps John's reaction was "Well, it depends on what the alternative is!" Or "What are the chances that you will find a way to get me out of this situation within another five years?" Or "Can you give me some time to think about it?" The possibilities are many, and any one of them would have yielded a confusing pattern of brain activity that we would find impossible to decipher, because John was neither imagining playing tennis nor moving through the rooms of his home—and these were the only two brain states that we could reliably interpret and understand. We were out of time. Melanie, Audrey, and Martin pulled John out of the scanner and sent him back to the ward.

≈

Communicating with John was even more thrilling than discovering that we could detect consciousness in vegetative patients. In John's case, he showed he had more going on cognitively than mere awareness of his surroundings. We'd even come close to answering one of the biggest questions of all—"Do you want to die?" Close, but not quite close enough.

You might think that answering questions such as whether you have any sisters is relatively easy for your brain to do, but it's actually quite complicated. Answer the question yourself. Do you have any sisters? I bet that felt easy. The answer doubtless came to you almost without thinking about it. Knowing whether you have any sisters comes easily because it's typically a situation that we have lived with for our whole lives. There are exceptions; maybe you had a sister, but she has passed away, making the question a little harder to answer without adding additional detail. But for most of us it's a simple yes or no. Yes, you do have a sister, or no, you don't.

But how does your brain do that? How does it know? The answer is, it doesn't just *know*, at least not in the sense that most of us feel that we, as people, just know certain things. Your brain can't just "know" that you have a sister any more than your computer can "just know" that you have a sister. It has to find out. Your brain has to search your memory for any evidence that you have a sister. That evidence can come in two general forms. It may be autobiographical, in the sense that you may have memories of growing up, playing with a person who looked a bit like you, and who answered to the same parents. Perhaps you remember your sister's twenty-first birthday and the present you bought her. That's an autobiographical memory that your brain can use to determine whether you have a sister.

The other kind of evidence that your brain may find is what psychologists call a declarative memory, or put more simply, *knowledge*. Somewhere in your brain a piece of data says that you do, or do not, have a sister. It has nothing to do with the experiences that you might have had *with* your sister; it's just a stored fact that you can pull out anytime you need to answer the question "Do you have any sisters?" It's a piece of knowledge, such as that Paris is the capital of France—knowledge that you probably know whether or not you've ever been to France. You learned that fact, just as you learned that you have a sister.

The distinction between autobiographical memory and declarative memory is of great interest to neuropsychologists because brain damage can affect one type of memory and not the other. In fact, my colleague Brian Levine at the Rotman Research Institute in Toronto has described a whole new condition known as severely deficient autobiographical memory syndrome, whereby the ability to vividly recollect past events is impaired, while other memory abilities are spared. These people might have no childhood recollection of their sister at all, no shared sibling experiences that they can report, no fond twenty-first-birthday memories. Yet they *know* that they have a sister because they have not lost the factual knowledge, the declarative memory for that information, and this allows them to lead more or less normal lives, their memory deficit often passing almost unnoticed even to themselves. Brian's cases typically have no history of brain injury or neuroimaging evidence of brain damage. So the root of the problem remains a complete mystery.

One conclusion that we could draw, then, is that John retained memories laid down before his accident, including where he had last been on vacation. Whether he used his autobiographical memory or his declarative memory we do not know, but one or both of these cognitive processes was intact, allowing him to answer the questions. And we were able to conclude a whole lot more about John's brain than that. Think about what else you need to do to answer the question "Do you have any sisters?" At the very least, you need to understand spoken language. If you don't understand the question, you certainly can't answer it. In addition, you need to hold that question in working memory for however long it takes your brain to retrieve the answer. What if you had no working memory, no ability to hold on to information until it's needed—in this case, to answer a simple question? Your brain would go off in search of an answer only to find that it had forgotten the question!

In fact, quite a lot more working memory was required for John to achieve what he did on that day because it wasn't only the questions he had to keep in mind. Throughout the scan, which lasted well over an hour, he had to remember what he had to do if the answer to a question was yes (imagine playing tennis) and what he had to do if the answer was no (imagine walking around his house). More important, John's responses that confirmed that these cognitive processes had to be intact also told us a lot about which parts of his brain were still functioning normally. If he could comprehend language, then the speech areas in his temporal lobe must have been working just fine. He could retain information in his working memory, which told us that the parts of his frontal lobe that are responsible for the highest forms of cognition were still responding as they should. He could also recall events from before his accident, which told us that the medial temporal lobe regions and the hippocampus deep within his brain were all still intact.

These mental processes are all things that you and I do routinely, moment to moment, without even thinking about them. But to witness this kind of elaborate scaffolding of consciousness announce itself in a patient who everyone had assumed was vegetative for five years was revelatory!

Though John could reliably and effectively "communicate" with us from within the scanner, Steven's team were unable to establish any form of communication whatsoever at his bedside. Communication via fMRI was all there was for John; it was the only option. Nevertheless, after the fMRI analysis was completed, a thorough retesting using standard neurological techniques led doctors to change his assessment to "minimally conscious." Somehow, knowing that John was in there must have made it easier for Steven's team to spot subtle signs of partial awareness, signs that had eluded detection before the scan.

John was only in Liège for a week. He'd been transferred from Eastern Europe for his assessment by Steven's group, and it was time for him to return home. We were out of time and out of luck. Many years later I asked Melanie what had become of him. After he returned home, Audrey had lost touch with the family. The phone numbers they had provided were disconnected, and there was no other way to make contact. John had disappeared as suddenly as he had appeared. After a few hours in the light, he was back in the gray zone, with no way to break out again.

These chance encounters with patients that came and went were frustrating, but it was a frequent occurrence back then. We were casting our net far and wide and sometimes transporting patients over great distances. Often, the logistics and the economics just got ahead of the science. As badly as any of us might have wanted to hold on to John, to explore his situation further, to delve even deeper into his inner world, it was impossible—we had to work with the circumstances, whatever they were. We were opportunistic wherever we could be, but frequently we were left disappointed. Science is often a random business, and progress frequently happens serendipitously, rather than through intelligent design. Nevertheless, it made me uncomfortable that we'd lost contact with John—I resolved to change things, to create a situation where we could follow patients indefinitely, regardless of their circumstances.

≈

When our paper describing John's case was published, my lab was once again deluged with frenzied media attention. My phone at the Unit wouldn't stop ringing. Camera crews came and went. I lost track of the number of times I appeared on some foreign radio station, recounting the story of the vegetative patient who could finally communicate with the outside world. The public seemed to have an insatiable appetite for the story, and the timing couldn't have been better. Martin was on the job market, and the very day he interviewed at UCLA, the *Los Angeles Times* ran with the headline "Brains of Vegetative Patients Show Life." It was no surprise that he got the job.

As has so often been the case, all of the attention influenced the science, and it influenced those of us whose careers depended on it. From our initial scan of Kate in 1997, when I had no funding at all to support this kind of research, to 2010, when John's story broke, the flow of money from grants and institutional support had changed considerably. The James S. McDonnell Foundation in the United States had awarded Niko Schiff, Steven Laureys, and me $3.8 million to develop a combined program of research. A group of us in Europe, including Steven, had picked up a grant worth almost 4 million euros ($4.5 million) to develop brain-computer interfaces for behaviorally nonresponsive patients, and the Medical Research Council had given me an extra £750,000 ($1 million) to extend our fMRI work in vegetative-state patients. Plus, much of my research program at the Unit was now focused on and funded for research on disorders of consciousness. In terms of research dollars, times were good.

With all of this attention came another game-changing hand. Out of nowhere, Canada came calling again. I was approached by Mel Goodale, a cognitive neuroscientist at the University of Western Ontario in Canada, famous for his work on visual perception and motor control. He told me about a recent scheme, initiated by the Canadian government, to bring foreign scientific "talent" into Canada. Successful candidates would be awarded $10 million in funding from the Canada Excellence Research Chairs (CERC) program, with matching funds from the host institution.

I seized the opportunity to move back across the Atlantic, start again from scratch, and set up Gray Zone II at Western's world-renowned Brain and Mind Institute, a new lab with better resources, better funding, and a whole new world of possibilities.

≈

Shortly after arriving in Canada, I got a call from a former colleague of mine, Dr. Christian Schwarzbauer, a physicist who was now working in Aberdeen, Scotland.

"We've been using your fMRI methods to scan patients who are in a vegetative state up here in Scotland," he said, "and we've recently scanned an old friend of yours." I immediately knew that he must be talking about Maureen. Her parents had made the connection between Christian and me and asked whether I would be prepared to comment on the results of her scans. Christian was also keen to seek my opinion.

It was the least I could do. But when it came to evaluating the scans, I was churning inside. I shut my office door: I needed solitude. Peering at the images of Maureen's brain felt like peering into the depths of my distant past. It was the strangest feeling—like touching some faraway emotional part of myself that I had buried years before. I was staring down at the brain of someone I had once been so close to. As I stared, I realized that the overwhelming animosity I had felt for our relationship had long gone. I was peering into Maureen's brain, looking for signs. Not of the person who had left me frustrated and confused, but the person I had once loved.

Christian had asked Maureen to imagine playing tennis and then to imagine walking through the rooms of her house. What was I going to do if her scan showed a response? I pushed the question to the back of my mind and peered once again at the screen in front of me. All I could see was darkness. A void. There was nothing there. Nothing of the Maureen I had once known. Nothing of Maureen at all. Ever elusive, ever unknowable—she was still a mystery.

CHAPTER TEN

ARE YOU IN PAIN?

≈

> It would be better to die once and for all than to suffer pain for all one's life.
>
> —Aeschylus

On December 20, 1999, a young man pulled away in his car from his grandfather's house in Sarnia, Ontario, with his girlfriend in the passenger seat beside him. Scott had studied physics at the University of Waterloo and had a promising career in robotics ahead of him. But at an intersection just a few blocks from his grandfather's house, a police cruiser traveling to the scene of a crime T-boned their car, hitting the driver's side full on. The police officer and Scott's girlfriend were taken to the hospital with minor injuries. Scott wasn't so lucky; his injuries were devastating. He was admitted to Sarnia General Hospital, and within hours his score on the Glasgow Coma Scale—a neurological scale that is used all over the world for measuring a person's conscious state—was rapidly dropping. Three indicators of awareness are rated: the eyes (from "does not open eyes" to "opens eyes spontaneously"), speech output, and motor responses. The lowest score possible is 3, indicating "does not open eyes," "makes no sounds," and "makes no movements." The highest score, 15, indicates that you are fully awake, conversing normally, and obeying commands. Scott was already a 4, just one step away from complete shutdown. Despite no outward signs of head or facial injury, his brain was badly beaten. The impact of the police cruiser into the side of Scott's car had slammed his brain against the inside of his skull, squeezing it into herniation and bruising it beyond recognition. Scott was in a bad way.

Twelve years later, soon after arriving in London, Ontario, I heard about Scott. I had contacted Bill Payne, a doctor at Parkwood Hospital, a long-term-care facility on the south side of the city, asking whether he knew of any patients who might be suitable for our studies. Originally established in 1894 as the Victoria Home for Incurables, Parkwood Hospital was still home to many "incurables" in practice, if not in name. Scott was first on Dr. Payne's list. "He's an interesting guy," Bill said. "His family are convinced he's aware, but we've seen no signs of it, and we've been observing him for years!"

I took a look at Scott. He certainly looked vegetative to me. But I needed an expert second opinion, and no one could provide a better opinion than Professor Bryan Young, a senior neurologist in the area. Approaching retirement with many years of experience with vegetative and comatose patients, he was possibly the nicest person you could ever meet.

I gave him a call: "What do you think of Scott?"

"Very interesting guy." This was starting to sound familiar. "His family are convinced he's aware, but we've seen no evidence of that."

I probed a little more deeply. Bryan had been seeing Scott regularly since his accident twelve years earlier. As the local neurologist with the most experience of disorders of consciousness, Bryan had naturally been the one who had examined Scott most closely. Bryan had immense experience by any measure, and an international reputation for meticulous and careful assessment of patients. If he thought Scott was vegetative, then I knew chances were that he was. I told Bryan that I was thinking of putting Scott into the fMRI scanner, and Bryan agreed that this was a good idea. "Please tell me what you find," he said.

I set off to Parkwood to assess Scott more thoroughly along with Davinia Fernández-Espejo, one of the postdocs who had moved with me to Canada from Europe. In a quiet room off the ward where Scott was staying, a nurse introduced us to his parents, Anne and Jim.

Anne, who had worked as a lab technologist, gave up work on the day of Scott's accident. Her husband, Jim, was a former banker and trucker. They were a lovely couple, clearly devoted to Scott and his life, such as it was, postinjury. Following the accident, they had relocated to a one-story bungalow outside London, Ontario, where Scott could stay when he wasn't being cared for full-time at Parkwood.

Jim and Anne told us that, despite his diagnosis, they believed that Scott, who loved listening to music from *The Phantom of the Opera* and *Les Misérables*, was responding to them.

"His face is expressive," Anne insisted. "He blinks. He does thumbs-up for positives."

Given Bryan's multiple assessments over the years, coupled with our own evaluation of Scott's condition, this was a curious comment indeed. We couldn't make Scott do thumbs-up no matter how hard we tried. I checked his official medical history. Neither Bryan nor any of the other doctors who had examined Scott over the years had indicated that he could do thumbs-up since his injury. Nevertheless, his family were adamant: Scott was responsive, and therefore Scott was aware.

≈

Curious as it was, I had seen this scenario many times over the years. A family is convinced that the person they love is aware in the absence of any clinical or scientific evidence to support it. The family speak and interact with that person as though he or she is fully conscious. Why? Do these families have some kind of heightened sensitivity to the patient's mental state? A kind of sixth sense for detecting consciousness that eludes even highly trained professionals such as Bryan Young? The family would certainly know the patient a whole lot better, which might explain their sensitivity to subtle signs of awareness.

One consequence of the brutality and abruptness of most serious brain injuries is that the doctor who assesses the patient—usually a trained neurologist—has generally not met the person in his or her former, healthy life. All the doctors "know" of the patients is what they see *after* the accidents. The family has the benefit of years of experience, a much more complete picture of the person within. Families also typically spend a lot more time with the patient after the accident. Neurologists, like all doctors, are busy and have a pile of clinical commitments and patients. That limits how much time they can devote to any one person. By contrast, many family members sit at the bedside for hour after hour, day after day, clutching to the faintest glimmer of hope, watching for the tiniest sign of awareness. It's natural that if it is there, they will be the first to see it.

But all that time, effort, and hoping is also sure to fuel wishful thinking, and the slightest hint of a response can alter a family's entire sense of reality. We're all terribly susceptible to what psychologists call confirmation bias, and confirmation bias is a real thorn in the side for gray-zone science. We tend to search for, interpret, favor, and recall information in a way that confirms our preexisting beliefs. If the person you love most is lying beside you in a hospital bed, her life hanging by a thread, you desperately want her to pull through. And you desperately want her to know that you're there. You ask her to squeeze your hand if she can hear you—and it happens! You feel a distinct increase in pressure as her hand gently squeezes yours. Your immediate reaction? She did what you asked, she responded, she's aware! It's a perfectly natural but unfortunately not scientific response. Science demands reproducibility.

Our world is chaotic; coincidences happen. Monkeys sometimes smile when we ask them to "Say cheese." Babies now and then point at the clock on the wall when we ask them, "Show us what time it is." And vegetative patients' hands occasionally tighten at that very moment when we ask them, in desperation, "Squeeze my hand if you can hear me." The result is intoxicating, almost magical. But are these tests reproducible? What if the next time you ask the person you love to squeeze your hand, she doesn't do it? Unfortunately, we are far less likely to take that negative response at face value. Therein lies the power of the confirmation bias.

Psychologists often use astrology as an example of the seductive power of confirmation bias. Why do so many intelligent and educated people believe, even a little bit, that the positions of the stars and the planets have some relationship to personality traits when no scientific evidence exists to support that? Psychologically, the reason seems to be that we attend more to information that matches what we already think than information about which we have no prior beliefs. When we meet people who are stubborn and then find out that they were born under the zodiac sign Taurus, a memory is activated in our brains—we are reminded that we "know" that Taureans are supposed to be a bit stubborn. So this (erroneous) belief is reinforced through reactivation. The problem is, when we meet another stubborn person who was not born under the sign Taurus, that memory—the link between the personality trait and the sign—is not activated. Nothing changes in our brain. We go on with our erroneous belief, neither stronger nor diminished.

To get rid of your erroneous belief, you have to start attending closely to all the stubborn people you know who were not born under the sign Taurus, as well as all the Taureans who are not stubborn. Eventually your brain would get the message that your belief has no basis in fact—a belief probably acquired when you were too young or too naive to appreciate the evidence.

This same skewed reasoning accounts for why many of us believe that redheads are hotheaded. Whenever we meet a fiery redhead, we notice it immediately because it confirms what we think. But all the calm and relaxed redheads pass by largely unnoticed. As a redhead, I am well aware that confirmation bias plays a role in prejudice: I have been accused of being hotheaded more than once by people who do not know me at all.

More broadly, confirmation bias also likely plays an important role in belief and faith. I remember attending our local Methodist church as a boy many years ago and listening to the minister applaud the efforts of a young girl who had recovered from life-threatening cancer. Throughout her ordeal she had attended church and the congregation had prayed hard for her. "Therein lies the power of prayer," opined the minister. I was troubled by memories of the many friends I had lost to cancer when I was in the hospital with Hodgkin's disease, some of whom had been just as religious and had had congregations praying just as hard for them. On balance the evidence shows that the "power of prayer" gives you, at best, an even chance. Nevertheless, our confirmation bias ensures that some of us go on believing in the face of an irrefutable mountain of contradictory evidence.

≈

As a scientist dealing with the families of patients who are in the gray zone, I have often found myself in the uncomfortable position of being privy to the most graphic and poignant examples of this very human tendency. When faced with a negative response, families will often make up reasons for why what they wanted to happen didn't. Perhaps the patient is tired now? Maybe the drugs have made her sleepy? Could it be she's in a bad mood and doesn't want to play the hand-squeezing game? Families cling to the one time a patient responded on cue to an instruction but ignore the countless other times that there was no response.

The power of confirmation bias is only half the problem. Imagine what happens when you are *not there* at the bedside. Imagine that hand squeezes occur regularly, all the time, with or without an explicit instruction to do so. It means nothing; like an itch that gets scratched, it's just a spontaneous automatic movement entirely devoid of conscious intent. When you arrive and ask the person you love to squeeze your hand, it happens! But when you leave and she is alone, it happens again. It has nothing to do with you, nothing to do with your instruction. But you are not there to know that. It's like a silent data point in time, exactly as important as the response that occurred when you were present, only lost forever because no one is there to witness it.

These two phenomena—confirmation bias and events occurring without witnesses—contribute to our tendency to place great weight on responses that we see and to completely disregard negative responses or responses that we don't see. Statistically, this is all data that should be given exactly the same weight.

I had no idea whether Scott's family had succumbed to a confirmation bias or whether they truly saw something in Scott that we could not measure. As a scientist I am prone to the former idea, but as a human being I am more than willing to accept the latter. It was impossible not to be moved by Scott's family and their utter devotion to making his life as comfortable as possible. I was also moved by their belief, whether scientifically valid or not, that he was aware. They were still there for him with an endless stream of support and belief in his ability to register the love they all so keenly felt for him more than a decade after his accident.

How could we not be swayed by their magnificent devotion? We tried many times, but we could never reproduce any kind of physical response from Scott under scientifically controlled conditions. We asked him to look at a mirror that we held up in front of him—nothing. We asked him to touch his nose—nothing. We asked him to stick out his tongue—nothing. We asked him to kick a ball—nothing. These are all carefully considered instructions that have been validated time and time again on hundreds of patients with serious brain injuries all over the world. It seemed to us that Bryan was right. The evidence suggested that Scott was indeed in a vegetative state.

≈

A BBC film crew had asked if they could record the scanning session with Scott, which added, for me at least, an extra level of anxiety to it. The BBC had been following our work for their series *Panorama*, which was first broadcast in 1953 and is the world's longest-running current-affairs documentary program. Our move to Canada had threatened to disrupt filming, which had begun in England, but in true British BBC spirit the crew decided to cross the Atlantic and follow our Canadian patients and the progress we'd made.

Medical correspondent Fergus Walsh hosted the show. I'd come to know him well because he was first on the scene in 2006 when we'd used fMRI to show that Carol was conscious, and he covered it generously on the BBC TV News. Fergus had also followed the Tony Bland case closely and came back to Cambridge in 2010 when we succeeded in communicating with a vegetative state patient for the very first time. But this was different—this was a one-hour BBC documentary to be broadcast around the world on prime-time television!

I was standing on the platform at the Cambridge railway station on a cold winter morning when Fergus first called me with this suggestion. The idea was to follow five patients from the point of their injury to final outcome, whether that turned out to be good or bad. Fergus hoped that at least one of these patients would be found to be conscious, and that, with luck, we could communicate with him or her.

I was skeptical: "It will never happen!"

"But you've claimed that up to one in five of your patients is conscious," Fergus insisted. "This is your big chance to prove yourself right!"

How can one not love Fergus? He's so enthusiastic—about everything, as far as I can tell. But he was putting me in a tough spot. We were going to be closely scrutinized by a BBC camera crew. What if we couldn't find another conscious patient? What if we couldn't again communicate with a nonresponsive patient? How would that look? Would people start to doubt what we'd been seeing and reporting? Would it undermine our whole research program? It felt risky. But that's par for the course. Much of science does feel rather risky and more than a little bit random. One year we'll see several conscious patients in a row, and the next year we'll see nothing for months on end. I thought back to Kate. We'd got lucky—she was one of the responsive ones. And we'd got lucky again with Carol. And with John. Could we do it again? On television? I had no choice; I had to try.

I agreed to be filmed, and Fergus and his team flew out to Ontario. A BBC camera crew followed me day and night. They filmed us in the lab. They filmed my band, Untidy Naked Dilemma, rehearsing in my basement at night. And they were filming Davinia and me the day we decided to scan Scott.

≈

As Scott lay in the scanner, Davinia and I went through the usual routine.

"Scott, please imagine playing tennis when you hear the instruction."

I still get goose bumps when I remember what happened next. Scott's brain exploded in an array of color—activation indicating that he was indeed responding to our request and imagining he was playing tennis.

"Now imagine walking around your house, please, Scott."

Again Scott's brain responded, demonstrating that he was there, inside, doing exactly what he was asked. Scott's family was right. He was aware of what was going on around him. He could respond! Perhaps not with his body in quite the way they had insisted he could, but with his brain! This fantastic moment was caught on camera by the BBC (www.intothegrayzone.com/mindreader).

What now? What should we ask Scott? Davinia and I looked at each other nervously. We badly wanted to push things to the next level, to ask Scott something that would be meaningful for him. Not something practical and bland such as whether he remembered his mother's name, but something that could, potentially, change his life. We had talked a lot about the benefits of asking a patient whether he or she was in physical pain. Pain is entirely subjective and can only be probed through self-report. We had already used our fMRI method to establish that Scott was conscious. Could we now use it to ask him whether he was in any pain? I tried to imagine what his answer might be. What if Scott said yes? The thought that he might have been in pain for twelve years was too horrible to contemplate. Yet it was a real possibility. If Scott said yes, he was in pain, I wasn't sure how I would respond. And then there was his family. How would they react? Suddenly, the presence of the BBC camera crew made the whole scenario a whole lot more complicated, but I couldn't change that. I had to go talk to Anne.

I held my head low to avoid the camera lens and asked Davinia in a hushed whisper, "Do you think we should do it?"

"We should. We *have* to."

I knew Davinia was right, we had to. Scott and his family deserved it. It was time to do something that might actually benefit one of our patients, time to do the right thing. If Scott was in pain, we needed to give him the opportunity to tell us that, and if so, we needed to do something to help him.

I stood up and walked slowly out of the windowless control room to where I knew Anne was waiting. The cameras followed me. Anne stood by the doorway, smiling.

My mind raced. "We'd like to ask Scott if he is in any pain, but I'd like your permission."

This was a pivotal moment. I was asking Anne whether we could, for the first time, ask a patient such as Scott a question that could potentially change his life forever. If Scott had been in pain for twelve years, no one would have known. It's impossible to imagine the endless nightmare that his life would have been.

We could have just gone ahead and asked, I suppose; but Anne was right there in the control room, and after everything she had been through, after all the years of hoping and thinking that Scott was in there, I knew I owed it to her to get some sense of whether that is what she wanted. I wanted *her* to be the one to say, "Do it!" And I wanted her to want this, for her and for Scott.

Anne looked up at me. Through this entire episode she had

remained stoic, almost cheery. I imagined that she must have come to terms with her son's situation many years earlier.

"Go ahead," said Anne. "Let Scott tell you."

I walked back into the scanning room, trailing the film crew. The atmosphere was electric. Everyone knew what the stakes were. We were going to push gray-zone science to the next level. This was no longer just a question of scientific progress—this was clinical benefit writ large. Again, thoughts of arguments with Maureen about the tensions between science for science's sake versus clinical care came flooding back like ghosts from my past.

"Scott, are you in any pain? Do any of your body parts hurt right now? Please imagine playing tennis if the answer is no."

I still shudder when I think about that moment (www.intothegrayzone.com/pain). We could barely breathe, leaning forward, backs straight in our chairs. Through the fMRI window we could see Scott's inert, mummylike body in the scanner's glistening hollow tube. The interfaces of multiple machines all worked together in elaborate synchronization so that our two minds could briefly touch each other and ask that most basic question: *Are you in pain?*

Davinia and I intently watched the screen. Fergus hovered silently over my shoulder. We'd come a long way since scanning Kate almost fifteen years earlier. Back then, we'd have to wait a week or more for the results to be analyzed. I could hardly believe that we used to sit by for a whole week waiting to see whether there'd been a response. In 2012, the results appeared on the computer screen before us more or less instantly. They were a whole lot sexier looking too. In 1997 our "results" consisted of a bunch of numbers on a page telling us where the activity was in the patient's brain, and whether it was statistically significant. By 2012 we had a three-dimensional structural reconstruction of the patient's brain—so lifelike you felt as if you could reach out and touch it. This brain image was the canvas on which "brain activity" in the form of brightly colored blobs was painted. They are beautiful images that vividly portray the brain at work.

As we peered at the screen before us, we could see all the folds and crevices of Scott's brain, both the healthy tissue and the tissue left irreparably damaged by the speeding police cruiser twelve years earlier. Then we began to notice something more. Scott's brain was springing to life, starting to activate. Bright red blobs began to appear; not randomly, but exactly where I was pressing my finger onto the computer screen.

Moments earlier I had said to Fergus, "If Scott is responding, we should see a response here," as I touched the shiny glass. And there it was. Scott was responding! He was answering the question! And more important, he was answering, "No."

There was a general collapse and congratulations throughout the room. Scott had told us, "No, I am not in pain."

I collected myself. I was close to tears. It was such a dizzying situation—a medical scientific breakthrough; the all-seeing eye of what would be prime-time watching; Scott's inert body lying motionless in the scanner; my team standing around in stunned wonder. The BBC film crew were beside themselves; they had got exactly what they wanted, but at that moment, for the first time in two years, I felt as if none of that mattered. This was Scott's moment, and he grabbed it. We could all see that.

After a few moments, the tension burst and everyone heaved a huge sigh of relief. Everyone, that is, except Anne.

When I told her the news, she was remarkably blasé. "I knew he wasn't in pain. If he was, he would have told me!"

I was a mess and could only nod my head dumbly. The courage of both of them overwhelmed me. She had stood by him all those years, insisting that he still mattered and deserved affection and attention. She had not given up on him. She would never give up.

Scott's response in the scanner simply confirmed what Anne

already knew. She knew Scott was in there. How she knew, I will never know. But she knew.

≈

The heart-wrenching moment when Scott told us that he was not in any pain became the centerpiece in what would be the prime-time BBC *Panorama* documentary "The Mind Reader—Unlocking My Voice." Watching it now, I can still feel the tension of that day in the scanner room. The program won awards and the reception was universally positive. But at the heart of it was something much more important than the publicity and the accolades. It had revealed a person, a living, breathing soul who had a life, attitudes, beliefs, memories, and *experiences*—who had the sense of being somebody who was alive and in the world no matter how strange and limited, at least outwardly, that world had become. For twelve years, Scott had remained silent, a silent somebody, locked inside his body, quietly watching the world go by. His mother had known he was there, intact and preserved; her son who was still her son.

On that day, and on many occasions in the months that followed, we conversed with Scott in the scanner. He expressed himself, speaking to us through this magical connection we had made between his mind and our machine. Somehow, Scott came back to life. He was able to tell us that he knew who he was; he knew where he was; and he knew how much time had passed since his accident. And thankfully, he confirmed that he wasn't in any pain.

The questions we asked Scott over the next few months were chosen with two goals in mind. In part, we tried to help him as best we could, by asking questions that might improve his quality of life. We asked him whether he liked watching hockey on TV. Prior to his accident, like many Canadians Scott had been a hockey fan, and so his family and carers would naturally tune his TV to a hockey game as often as they could. But more than a decade had passed since Scott's accident. Perhaps he no longer liked hockey? Perhaps he'd watched so much hockey that he couldn't stand it any longer? If so, checking in to see what his current viewing preferences were might significantly improve his quality of life. Fortunately, Scott still enjoyed watching hockey, much as he had for many of the years prior to his accident.

I have seen this scenario countless times with different patients; choices are made about leisure activities based largely on what they enjoyed prior to their brain injury. As a patient, if you enjoyed heavy-metal music, that's what you'll get to listen to as you while away the hours in your hospital bed. The problem is, many years may have passed, patients may have grown from adolescence into adulthood in their hospital beds, but the music doesn't change. It's as if time stands still.

I heard one story about a patient who loved the Canadian artist Celine Dion. But she only owned one Celine Dion album. Fortunately for her she recovered, and when she did, her first words to her mother were "If I ever hear that Celine Dion album again, I will kill you!" Hours and hours of listening to Celine Dion would jeopardize anyone's quality of life, but imagine you were confined to bed and could do nothing to stop it. A recipe for going quietly insane.

The second type of questions we asked Scott were chosen to reveal as much as possible about his situation, what he knew, how much he remembered, what sort of awareness he had. These questions were less about Scott the person and more about our digging deeper into the gray zone. Understanding what situations were psychologically possible in this limbo was incredibly important because no one knew the answers, and as it turned out, many people had made wildly erroneous assumptions.

For example, after lecturing about patients in the gray zone, I'd often heard comments like "Well, I doubt they have any sense of the passage of time" or "They probably don't remember anything about their accident." Or even "I doubt they have any awareness of the predicament they're in."

Scott told us otherwise. He answered all of those questions and more. When we asked him what year it was, he told us correctly that it was 2012, not 1999, the year of his accident. Clearly he had a good sense of the passage of time. He knew that he was in a hospital and that his name was Scott. Clearly he had a good sense of who he was and where he was. Scott was also able to tell us the name of his primary caregiver. This was important to us and to our understanding of gray-zone science because one question that had frequently come up was what patients in this situation could remember. Scott would not have known his caregiver prior to his accident, so his knowing her name was clear evidence that he was still able to lay down memories.

Laying down memories is central to our sense of time passing, of life moving along, of our place in the ongoing scheme of things. Imagine that every day you woke up and could recall nothing that had happened since the day you had an accident, say ten years earlier. How would things feel? Your nurse, who may have cared for you day and night for a decade, would seem like a complete stranger. Your family and friends, whom you recall well from before your accident, would suddenly all look ten years older. And your home, assuming you still lived in the same place, would feel as if it had been completely revamped overnight—every change that had occurred in the interim, every wall that had been painted, every piece of furniture that had been moved or replaced, all of these changes would seem to have occurred in the few hours since you went to sleep.

Worse, if you'd moved since your accident, you would have

no idea where you were at all! A condition known as anterograde amnesia is a lot like that. Patients with anterograde amnesia are typically unable to lay down new memories, while their "old memories"—those laid down before the onset of their amnesia—remain largely intact. The most famous case of anterograde amnesia is Henry Molaison, or H.M., as he is more widely known. In 1953, H.M. underwent surgery to try to fix his persistent seizures—his hippocampus and the cortex around it on the inside surface of the temporal lobe were removed on both sides of the brain. As a result, Henry became unable to remember anything new that happened to him, despite being able to recall events from his childhood just fine. Much of what we know about the role that the hippocampus and the surrounding brain areas play in memory can be traced back to H.M.'s rather unfortunate, but necessary, surgery.

In the UK another remarkable case of anterograde amnesia is Clive Wearing, who was, until March 1985, enjoying a successful career as an expert on early music with BBC Radio. He then contracted a herpes simplex virus that attacked his brain. His hippocampus was damaged, and since his brain injury, he has been unable to store any new memories for more than about half a minute. He spends each day "reawakening" every twenty seconds or so, "restarting" his stream of consciousness. He has lost any sense of where he is in the passage of time. He greets his wife joyously whenever they meet, even though she may have left the room only a few minutes earlier. Clive often reports feeling as if he has just woken from a coma. He is living constantly in the moment, like an island of consciousness moving through time, completely unaware that the world is changing around him. It's a nightmare scenario, yet paradoxically, his condition spares him any complete understanding of his plight.

Because of cases such as Henry Molaison and Clive Wearing, we felt it was important to establish that Scott's experience of life was not that of an island of consciousness moving through time. We felt that it was critical to know not only that he remembered his past, but that he was aware of the present and aware that today's present would be tomorrow's past. We wanted to know that Scott had the experience of existing in time, of being here today as part of an evolving history with events that come and go, all influencing and being influenced by other events on that same timeline.

≈

Throughout Scott's returning to the scanning center time and time again to be asked questions about life in the gray zone, his mother, Anne, remained cheerful and supportive. Clearly, not all of these trips were for Scott; some were for science. In a fine balance, we juggled questions that might be useful in improving his life with questions that might be useful in understanding and perhaps improving the lives of the many other patients in the gray zone.

Anne seemed to understand that. I wondered whether her former life as a lab technologist had taught her about this balance, between what's good for the patient and what's good for science. I never asked.

≈

Scott died in September 2013 of medical complications from his original accident. This is an all-too-common outcome, even many years after a serious brain injury. All that lying around and exposure to the army of obnoxious viruses, bacteria, and fungi that populate every hospital ward deadens the immune system and makes you highly susceptible to conditions such as pneumonia. After several weeks fighting infections, Scott died at Parkwood.

It shocked my whole team. We had spent many hours with Scott and he was part of the family. We had never had a real conversation with him, yet bizarrely we all felt we knew him. He had touched us deeply. We had dug deep into his life in the gray zone, and he had responded with answers that left us in awe of his strength and courage. His life had become interwoven with ours.

At the wake, it was lovely to see Anne and Jim again, although I wish the circumstances had been different. The funeral home was packed. Scott's body lay in an open coffin toward the back of the room. Friends and family had come from near and far. Despite his fourteen years of being mostly inside himself and cut off from the world, at the time that Scott died many people still felt a profound connection with him.

Jim asked if I would like to see Scott. I was taken aback. I have attended many funerals, but in the UK, where I am from, open coffins are rare, and I had never had that experience. I wasn't sure what to do. But I had enormous respect for Jim and the rest of Scott's family, and so I went to see Scott one last time.

I had such an odd response. In many ways, Scott looked as he had always looked to me. I hadn't known the real Scott, the Scott who had lived a full and happy life, who walked and talked and laughed and moved purposefully through the world until the age of twenty-six, when all that was suddenly and permanently taken from him. I had only known this Scott, the physically nonresponsive Scott, the Scott lying in front of me right now. It occurred to me right then that this gray zone, this place that is home for many of our patients, truly is the borderland between life and death. It's so close to death that sometimes it's hard to tell the difference between the two. Scott was still there in the way that he had always been there for me, even though now he wasn't there at all.

≈

On Scott's obituary web page I wrote, "It was a great privilege getting to know Scott these past few years. His heroic efforts for science will never be forgotten and will be reflected in the lives and minds of all of us who knew him and many more who didn't."

Fergus Walsh wrote, "It was a privilege to meet Scott—he was a remarkable and determined man. The reports we did about Scott's abilities to communicate despite his disabilities were seen and heard around the world. Sincere condolences from all the BBC team to Anne and Jim."

The relationship that we developed with Scott and his family was unlike any other that my team has experienced before or since. In part it was Anne and Jim's warmth and openness in sharing their world with us and bringing us into their lives, but more than that, Scott himself created and sealed our bond. To communicate for the first time with another human being who has been unable to communicate for more than a decade is an extraordinary experience. To do it again and again is magical. Scott let us into his world, and we laughed with him, joked with him, and cried with him. When that door shut and Scott was finally gone, I think a little part of all of us died with him.

CHAPTER ELEVEN

LIVE OR LET DIE?

≈

That I might drink, and leave the world unseen,
And with thee fade away into the forest dim

—John Keats

Scott's death reminded me how dangerous modern life is. He was killed, ultimately, by a speeding police cruiser, but it took him fourteen years to die. Driving is dangerous. Around thirty-seven thousand people die on American roads each year. For every death, many more do not die, at least not at the roadside. Some slip into the gray zone, languishing until they expire. But why does this happen? How do they get there? Why don't they recover? Why don't they die immediately? How do they end up in that awful in-between place?

Fifteen years into probing the edges of the gray zone, I still had no answers to these questions. Why does the brain shut down sometimes and not others? Are some of us more innately resilient? Is one part of the brain the culprit? If so, which part?

Our explorations into the gray zone had provided more questions than answers. We had learned that many doors lead to the gray zone. A common route is when what might be called the window of opportunity is missed. When a patient arrives in a hospital after a serious brain injury, for some period, usually days or a few weeks, the prognosis—the likelihood of making a reasonable recovery—is *completely* uncertain. That's because every brain injury is different.

During this period, patients are usually on life support. They are likely to be intubated—a flexible plastic tube is inserted into the windpipe through a hole in their neck to assist with breathing. They may well be on a ventilator, which keeps oxygen flowing through the body by pushing air into and out of the lungs. Before these amazing technologies existed, you would just have a serious brain injury and then die. But machines improved your chances, shepherding you through those crucial first few days. And some people do, indeed, survive. Their bodies reboot, but their brains do not. Not fully at least. We created the gray zone, or at least we massively increased the possibility that any of us could survive in it.

People have always entered the gray zone, but they probably did not survive there for long. After receiving a blow to the head, a prehistoric human would likely have been "knocked out" in much the same way that Muhammad Ali dispensed with most of those who dared to take him on. Like many unfortunate boxers, if that state of unconsciousness persisted for more than a few minutes, then a "coma" may have ensued—a prolonged failure to respond to any form of stimulation, an absence of normal sleep-wake cycles, and a failure to initiate any voluntary actions. Without modern medicine, the chances of a prehistoric human emerging from his or her comatose state would have been low; apart from anything else, without nutrition and hydration he or she would likely have deteriorated rapidly and quickly died. Indeed, the chances of surviving a prolonged coma are still not high; for patients, such as Scott, who enter the ER with a Glasgow Coma Scale score of 4 and receive all the help that modern medicine can provide, 87 percent will either die or remain in a vegetative state forever. The chances of a prehistoric human surviving this period and squeaking through into the gray zone were negligible.

Nevertheless, people did exist in the gray zone before the advent of the artificial ventilator in the 1950s. The ancient Greeks referred to a condition they called apoplexy, which sounds eerily similar to what we now call the vegetative state: "The healthy subject is taken with sudden pain; he immediately loses his speech and rattles his throat. His mouth gapes and if one calls him or stirs him he only groans but understands nothing. He urinates copiously without being aware of it. If fever does not supervene, he succumbs in seven days, but if it does he usually recovers."

Between ancient Greece and the twentieth century not much changed in our understanding, diagnosis, and treatment of patients presenting with this peculiar behavior. In the middle years of the twentieth century other descriptive terms started to be used, including *coma vigil*, *akinetic mutism*, *silent immobility*, *apallic syndrome*, and *severe traumatic dementia*. Whether these terms described the same or different conditions is entirely unclear because (as is still the case today) every patient was different and therefore the precise pattern of symptoms varied widely. This probably explains why none of these terms was universally adopted. The term *pie vegetative* was used in 1963 and *vegetative survival* in 1971, predating the introduction of *persistent vegetative state* in a landmark paper published by Bryan Jennett and Fred Plum in the *Lancet* on April Fools' Day in 1972. It quickly entered common medical parlance.

≈

For the modern patient entering the neurointensive care unit, if the test results are poor and indicate that the patient may soon die or will never recover to any form of life worth living, the family may be advised to "withdraw life support"—turn off the ventilator, or "pull the plug" as it is often colloquially known. For some families, perhaps those with blind faith in the medical establishment, or those who know that the best-case scenario is not what their loved one would ever have wanted, agreement comes easily and they pull the plug with little hesitation.

For others, however, the decision is much harder, and they agonize for days and days. And therein lies the problem: if, during this "window of opportunity," patients recover to the extent that their lives becomes self-sustaining—they no longer need the ventilator to keep them alive—then the opportunity to pull the plug is lost. They have reached the gray zone, and no longer can a simple plug be pulled to end their lives. It can only be ended by withholding food and water.

This fine but important legal distinction revolves around whether we think of food and water as "medical treatment." Ventilators are clearly medical treatment, and decisions to turn them off are made relatively easily in some cases (for example, when there is no chance of recovery). But is food and water a medical treatment? Some jurisdictions think it is, and others consider it to be a basic necessity, or right, that cannot be withheld. One factor that undoubtedly influences opinions is how long it takes to die by the processes. After a ventilator is turned off, a patient will typically die in minutes due to the lack of oxygen to the brain. By withdrawing nutrition and hydration you are starving the patient to death, which can take up to two weeks.

The change from pulling the plug to withholding food and water is relatively subtle but crucially important in the minds of philosophers, ethicists, and lawyers. Now the family has to decide not whether to keep the patient alive, but whether to help her *die*.

At the Royal Society in London, I recently co-organized a meeting on consciousness and the brain with my friend and colleague Mel Goodale. The topic of the meeting was how best to measure consciousness, and we had many great thinkers in attendance, including philosophers, cognitive neuroscientists, anesthetists, and robotics engineers. As we sat around pondering how best to measure consciousness, the debate shifted into a lively discussion of how our sense of consciousness and humanity influences the ease with which we kill, which seems to be closely related to the physical form and the behavior of what is being killed and its similarity, or not, to human form and behavior.

Think about boiling mussels. Relatively few people are troubled by throwing a bag of them into boiling water. By any standard, this is a fairly brutal way of ending the life of a living creature. But mussels are not much like human beings. They have no arms or legs or discernible humanlike features. They do not behave as we do, moving around all over the place and interacting actively with their environment.

Now consider a lobster. This is a harder problem. Many people are squeamish about boiling lobsters alive, preferring to buy them precooked from the store. Lobsters are also not much like humans, but they are a whole lot more like humans than mussels. They have legs and appendages that resemble, functionally at least, human arms. Like us, they grip things. They have eyes, and if you study a lobster it's easy to convince yourself that, unlike mussels, they have some sort of face. Lobsters also move around their environment, interacting with it in ways that, while obviously very different from those of humans, certainly resemble some of our behaviors.

I won't take this line of thought much further; suffice it to say that I am quite confident that few of us would be comfortable throwing a monkey or an ape into boiling water. Why? Why are we so much more ready to boil a mussel to death than a lobster? Clearly, a spectrum of behavior around physical form drives our feelings about what we are doing when we boil a mussel or a lobster—an identical action, except that these are two different types of shellfish.

At the heart of these feelings, I believe, is the sense we have about how *conscious* each of these living creatures is. A lobster is probably a bit more conscious than a mussel because it is a bit more like us than a mussel. But do we have any evidence for that? As we have seen previously, our assumptions about consciousness are largely based on behavior rather than some established biological facts. Even if there is scientific evidence that lobsters are more "conscious" than mussels, I doubt many of us have read the relevant academic papers to support that, preferring instead to make the decision based on our intuitions.

But where is the threshold—the evolutionary threshold, if you like—that determines whether we think another creature is conscious? If we mostly think that mussels are not conscious, and we mostly think that monkeys and apes are conscious, somewhere in between those two places, consciousness (or at least our strong intuitions about consciousness) must emerge. That some of us are willing to boil lobsters while others are not suggests to me that lobsters are somewhere near that critical threshold. Most of us, however, don't think that mussels are conscious, so far fewer of us have a problem boiling them to death.

≈

This critical spectrum between the unconscious and the conscious is central to many of the agonizing decisions that families have to make at the bedside. Lying in a hospital bed in an intensive care unit, patients rarely behave as we do. They often don't move and rarely show signs of interacting with their environment. Although not really *like a* mussel, they are behaviorally more like a mussel than they were before their brain injury. Physically, many of the cardinal human features that we know and love are often grotesquely altered: faces disfigured, limbs irreparably damaged, twisted, or missing altogether.

These factors undoubtedly drive our assumptions about consciousness (as they do with nonhuman creatures). If patients don't behave like humans and no longer even look human, then it's much easier to believe that they don't think like a human either. In turn, these factors contribute to how easy or difficult it is for us to decide whether a person we love should live or die. Would it be harder to pull the plug on a physically preserved patient than on one whose body has been battered beyond recognition? Why? On one of the occasions that Maureen's brother Phil and I met up, he told me that the family had agonized for many years about whether it was better to treat every infection, or to let her succumb naturally, as often occurs in cases like hers. Whether this decision was made more difficult by her remarkably preserved physical condition I do not know, but I am sure it didn't make it any easier.

We also know that it is harder to decide to end a life if a patient has missed the window of opportunity and entered the gray-zone vegetative state—seemingly awake but unaware. And it is all but impossible to end a life if some physical response, even one as subtle as the blink of an eye, indicates that someone, a person, is inside a nonresponsive body.

Our willingness to end life is inextricably linked to our assumptions about what life means, how much of "us" is left when the dust has settled following a serious brain injury. But as we now know, this is folly—how much of a person remains often has little to do with what we see lying before us.

≈

Abraham was in his sixties when he had a major stroke in 2014. His wife had brought him into the emergency room after he developed a sudden headache, started vomiting, and appeared confused and disoriented. A CT scan revealed that he had sustained a large intraventricular hemorrhage—bleeding into the fluid-filled cavities (or ventricles) deep within the brain. He was immediately sedated, intubated, and moved to the intensive care unit. Further scans revealed that an aneurysm, or a weakening in the wall, of the anterior communicating artery (so called because it connects two major arteries, one in the left and one in the right hemispheres of the brain) had ruptured, resulting in severe damage to the surrounding area, including Abraham's left frontal lobe.

At the time that we scanned Abraham, twenty-two days after his stroke, he was comatose but moving toward vegetative state. His eyes would intermittently open, and he was starting to take some breaths on his own. He was tall and I noticed his toes almost protruding beyond the end of the hospital bed.

It was an important day for our lab. My graduate student Loretta Norton was doing something completely new, attempting to scan patients who were still in the intensive care unit in *the first few days* after a brain injury. These were not medically stable patients, such as the ones we had been scanning starting in 1997 with Kate—patients who were typically months or even years removed from their life-changing injuries. These patients were clinging to life by a thread, their lives measured in hours and days rather than weeks and months. If we could find new ways to improve the diagnosis of these patients and even improve the accuracy of predictions about who was likely to die and who was most likely to survive, it would be a major advance for intensive care medicine. The ethics committee gave us permission for this pioneering study to scan this extremely vulnerable population, despite the risks.

Unusually, in my experience, Abraham had made it clear to his wife what he would want if he was ever on life support. He hadn't written an advanced directive, a legal document that sets out your wishes about medical care if and when you are incapacitated and unable to communicate them yourself; but he and his wife had discussed the issue in detail, and there was no question about where he stood. Abraham had clearly stated that he would never want to be kept alive in a vegetative state, and his wife relayed these wishes to the care staff and the doctors when they admitted him. His wife moved to act on his prior instructions. Discussions were started about how and when Abraham would be allowed to die.

When such decisions are to be made, a team of professionals meets with the family to ensure that they understand the issues. The team often consists of the most responsible physician (often a neurologist) plus a more junior medical resident, nurse, and social worker. After considering all the options, if the family agrees to withdraw life support, a time is set, usually within twelve to twenty-four hours, although it may be delayed to allow family members or friends to gather. Occasionally, if everyone is present, it happens immediately. The procedure is explained to the family, and they are generally allowed to be with the patient throughout it. The physician prescribes a cocktail of pain medication—sometimes referred to as comfort care because if it's not given the patient will usually appear uncomfortable, often gasping for air. Once comfort care is administered, the physician either shuts the ventilator down gradually or immediately (every doctor does this a little differently). Neither the pain medication nor the removal of the ventilator prevents the patient breathing on his or her own, and often this happens—usually for a short time but sometimes for many hours. Death is an unpredictable business.

Unfortunately for Abraham, he and his wife had been active in a church that held strong views about the sanctity of life. Indeed, Abraham's pastor, who had been an ever-present figure in the intensive care unit, declared that it was "God's will" that Abraham be kept alive. Abraham may have had a clear idea about where he wanted to end up, but according to this pastor, God had other plans for him. The final decision in such cases as this rests with the substitute decision maker, in this case Abraham's wife. I was shocked and a little disturbed when Abraham's wife decided that, despite his specific instructions, he should not be allowed to die.

"I have already lost my husband," she said. "If I don't do what my pastor says, I will lose my church as well."

≈

Complicated circumstances bring complicated decisions; with life, death, and the gray zone, these decisions often come with enormous ethical and moral consequences. In my experience, no two sets of circumstances are the same. In Terri Schiavo's case, disagreement between her husband and her parents about what Terri would have wanted fueled a national spectacle and more or less wrote the history books on how these cases would be dealt with in the United States. Here in Canada, with Abraham, we had our own little Schiavo incident, but the circumstances were different: a pastor and the "word of God" versus a wife and her future with or without the support of the church.

These two cases are equally troubling to me. As one who does not believe in rule-bestowing higher powers, I find making decisions based on the "word of God" to be nonrational—one might just as well make decisions based on the roll of a die. Yet, at the heart of it, I do understand the predicament that Abraham's wife was in. For example, if you substitute "her closest friends" for "the church," Abraham's wife's dilemma starts to make a bit more sense, and the complexities of religious persuasion have been entirely removed. If her closest friends threatened to disown her if she sanctioned her husband's death, perhaps her social life and best sources of support in the wake of losing her husband would be irreparably damaged. Unlike with Schiavo, however, Abraham's prior wishes were clear—he did not want to go on living in his current state. In my opinion, this trumps all else, including our ongoing relationship with our closest friends—our personal wishes take precedence even if they do not align perfectly with the wishes of those we leave behind.

Then again, every case is different. I was recently involved with a lawsuit concerning a fifty-six-year-old Canadian man. The patient, let's call him Keith, and his wife and their three children were in a serious car crash in September 2005 when Keith was forty-nine. The eldest son was instantly killed. Keith suffered a profound and irreversible brain injury. The wife and the two younger children suffered physically and emotionally, but emerged relatively intact. Keith was diagnosed as being in a vegetative state, and by 2012 his wife felt it was time to say good-bye. She instructed Keith's caregivers to remove his feeding and hydration tubes, which would lead, within a few days, to his death. Keith's brothers and sisters strongly opposed this move and petitioned the local court to prevent her from doing this. They also requested that Keith's wife be removed as the substitute decision maker (presumably so that she couldn't make a similar request in the future), and they be given this role.

The judge eventually dismissed these requests based on careful consideration of the circumstances. Keith and his wife had been married for twelve years prior to the accident and had three children—it therefore seemed perfectly reasonable that she was the substitute decision maker for Keith and, presumably, had his best interests at heart. This makes perfect sense to me. Substitute decision makers are generally substitute decision makers for good reasons, and it would be odd if another person or organization could overthrow that responsibility just because they had a different opinion about what should happen to a person who could no longer make his or her wishes known.

Unfortunately, it's not always that simple. Another recent case that went all the way to the Supreme Court of Canada was that of Hassan Rasouli, a sixty-one-year-old Iranian engineer who emigrated to Toronto with his wife and two children in 2010. That October he had surgery to remove a benign brain tumor and contracted an infection that left him seriously brain damaged. His doctors determined there was no hope of recovery and that keeping him on life support was futile; it would inevitably result in a series of progressively worse medical complications, infections, and so on that would require expensive treatment. For what? They recommended withdrawing life support. The patient's wife, Parichehr Salasel, who was the substitute decision maker, refused to consent, citing the couple's Shia religion and her belief that her husband's movements indicated some level of minimal consciousness.

As Canada's national newspapers reported, we had scanned Hassan some months earlier, and our fMRI scans also suggested minimal consciousness—he appeared to be able to imagine playing tennis and walk through the rooms of his home, although not consistently. When we assessed him behaviorally, the story was much the same; he could follow a mirror with his eyes and would fix his gaze on a family photo when it was held up in front of him, although again these responses were inconsistent. Nevertheless, in all likelihood, and in the opinion of seasoned medical experts, he had no chance of any significant recovery and would continue to be a significant drain on the Canadian medical system. The ultimate ruling, handed down by the Supreme Court, was that doctors cannot unilaterally decide to withdraw life support without the consent of the patient, his family members, or a substitute decision maker. Even if the doctors were right, even if their opinion was "in the best interests of the patient," even medical professionals with years of relevant experience can't overrule the opinions of the substitute decision maker—at least not in Canada. This is such a new field that the laws in different parts of the world are being made case by case.

≈

When it comes to the right to live versus the right to die, the United States has had more than its fair share of controversy. Aside from Terry Schiavo, two other prominent cases have profoundly influenced the legal and ethical issues surrounding our "right to die."

In 1975, Karen Ann Quinlan, of Scranton, Pennsylvania, went to a friend's birthday party at a local bar in New Jersey and had a few glasses of hard liquor and consumed some methaqualone (quaaludes). Karen Ann had been dieting and had eaten nothing for several days. Sometime later she reported feeling faint and was taken home and put to bed. When friends found her, she had stopped breathing. An ambulance was called and she was admitted to a hospital in a coma.

Karen Ann's parents, Joseph and Julia Quinlan, asked medical staff to disconnect her from her ventilator—she would often thrash around violently, and her parents believed that the ventilator was causing her pain. The doctors refused, fearing homicide charges would be brought against them if they complied with the Quinlans' request. The parents filed suit to disconnect Quinlan from her ventilator, arguing that it constituted an extraordinary means of prolonging her life. In court, the lawyer for the Quinlans argued that Karen Ann's right to die supplanted the state's right to keep her alive, while her court-appointed guardian argued that disconnecting her ventilator would be homicide. The judge ruled against the Quinlans. Following an appeal to the New Jersey Supreme Court, the Quinlans' wish was finally granted, and Karen Ann Quinlan was removed from her ventilator. What followed was unexpected and unfortunate. Karen Ann began breathing unassisted and lived for another nine years in a local nursing home, kept alive by a feeding tube, which her parents had not sought to have removed because, unlike the ventilator, they did not consider it to be an "extraordinary means for prolonging life." Karen Ann Quinlan died from respiratory failure in 1985. In many ways, her case marks the beginning of the right-to-die movement in the United States and continues to be discussed by law courts, ethics committees, and philosophers to this day.

≈

Another significant American case was that of Nancy Cruzan, who was twenty-five years old in 1983 when she lost control of her car and ended up facedown in a ditch full of water. After three weeks in a coma she was declared vegetative and a feeding tube was inserted. Five years later, her parents asked for her feeding tube to be removed, but the hospital refused on the grounds that this would lead to her certain death. In the year before her accident, Nancy had told a friend that if she was ever sick or injured, she would not wish to continue her life unless she could live at least halfway normally, and on that basis, the courts granted the Cruzans their wish. But in a counterpoint to the case of Karen Ann Quinlan, the Supreme Court of Missouri reversed the trial court's decision, ruling that no one may refuse treatment on behalf of another person when an adequate living will is not present.

Nancy's case eventually reached the US Supreme Court, which in a 5–4 decision favored the Supreme Court of Missouri. The US Supreme Court ruled that nothing in the US Constitution prevented the State of Missouri from requiring "clear and convincing evidence" before terminating life-supporting treatment. In cases such as Nancy's, the Court ruled that "clear and convincing evidence" was required because family members might not always make decisions that the patient would have agreed with, and those decisions (such as the withdrawal of life support) might have irreversible consequences.

In response to this ruling, the Cruzans gathered as much evidence as they could that Nancy would have wanted her life support terminated, given the circumstances, and convinced a local county judge to rule that they had met the evidentiary criterion of "clear and convincing evidence." In 1990, just before Christmas, Nancy's feeding tube was removed according to the local judge's ruling. In another bizarre turn of events, in the days following the removal of the tube, nineteen representatives of the right-to-life movement entered her hospital room and tried to reconnect her feeding tube (they were all arrested). Finally, on the day after Christmas 1990, Nancy Cruzan died. Six years later her father committed suicide.

≈

As shocking as these cases are, they illustrate both the complexity of the legal issues involved and the profound impact that severe brain injury has, not only on the victims' families, but also on society in general. Like Terri Schiavo, both Quinlan and Cruzan polarized an entire nation. They raise important questions about the legal differences between refusal of treatment, suicide, assisted suicide, physician-assisted suicide, and "leaving someone to die." What role does government have in such decisions? Should they be dictated by those closest to the patient, the attending physician, or government officials who may have their own biases about life, death, and everything in between? Or should we rely solely on advanced directives or "living wills" from the patient? If so, what should we do when no such directives exist? For some, the cases of Karen Ann Quinlan and Nancy Cruzan were high-water marks in the right-to-die versus right-to-life debate. For others, they were a step too far on the slippery slope to out-and-out murder.

≈

To return to Keith, the Canadian man who, with his family, was in a car crash—my involvement in his case came when his siblings, having read about my work, requested an expert opinion on whether Keith could be brought to London, Ontario, for an fMRI scan, and if so, whether it might show that he was conscious. Could it even be used, they wondered, to ask Keith what he wanted to happen?

Imagine that we had brought Keith to London, and imagine that we had put him into our fMRI scanner. Imagine further that we had found out that he was conscious and that he could answer yes-and-no questions. We had already shown that was technically possible. Keith was relatively young and healthy at the time he sustained a traumatic brain injury—all factors that we now know contribute to a positive fMRI response. Chances were good that Keith was conscious and also that he would be able to communicate that. What if Keith, with the assistance of our fMRI scanner, had told us that, contrary to his wife's opinion, he wanted to *live*? What if Abraham had been able to come to his wife's defense and confirm to the pastor that he wanted to *die*?

You would think that if someone with severe brain damage and thought to be in a vegetative state was suddenly able to tell you that he or she wanted to die, that's what should be allowed to happen. Shouldn't someone in that position have a clear right to die? The answer, I'm sorry to say, is not so cut-and-dried.

If an otherwise healthy person walked up to you and announced that he wanted to die, wouldn't your first reaction be to question his sanity? Or if not his sanity, at least his current state of mind? Perhaps he is simply depressed and unable to make a reasoned decision. And even if you could confirm that he was of sound mind, wouldn't you want to check the following day, and the following week, that he hadn't reconsidered? Perhaps he was just having some rotten days. With time, his extreme morbidity might pass.

Even if it persisted, even if you were a doctor and a patient came back to you day after day, week after week, announcing that he wanted to die, what could you do about it? The answer is nothing. Most of us do not live in a society that allows us to agree to, and assist in, suicide. Why should this be any different for a patient who has sustained a serious brain injury? Replying yes to "Do you want to die?" could reflect some kind of underlying psychological or psychiatric instability. The death wish might be transient. Would it still be there tomorrow or a year from now?

In any case, why should society allow us more freedom to pull the plug just because a person is in the gray zone? Should anyone be allowed to decide to die just because he or she wishes to? As a society we generally say no, but the technology is now there to allow people in the gray zone to make their own decisions about whether to go on living. At the very least, now that we know that many of them are not what they appear to be, we should all think carefully before making that decision on behalf of another.

Steven Laureys and his colleagues have conducted a study that suggests that what we all think we want to happen to us ("Please don't let me live in a gray-zone state") is not what we actually want when disaster strikes. His team surveyed ninety-one people with locked-in syndrome—conscious people who were only able to communicate by blinking or vertically moving their eyes. They were asked to answer a questionnaire about their medical history, current status, and attitude to end-of-life issues. Their quality of life was also assessed using a scale that ranged from +5 (well-being equivalent to the best period of their life prior to being locked in), to -5 (well-being equivalent to the worst period in their life ever). Contrary to what most of us might expect, a significant proportion of the patients (72 percent of those who responded) reported that they were happy. What's more, a longer time in a locked-in syndrome was correlated with how happy this group said they were!

While most of us declare that we would not want to live if we were "locked in" following a brain injury, only 7 percent of the entire group surveyed by Laureys and his team expressed a wish for euthanasia, suggesting that our preconceived notions about what we might think if the worst was to happen are false. On the contrary, most locked-in patients are reasonably satisfied with their quality of life—death is not the most frequent choice of those who have actually lived through this experience.

Studies such as this are obviously imperfect. Only 91 of 168 patients who were originally approached responded, and many of those who were most dissatisfied with their lives might have chosen not to complete the questionnaire. This is "selection bias," meaning that, for some reason, your sample is not representative of the entire population that you're gauging. Selection bias can produce misleading results.

Still, this is the best study available, and the data show that a significant proportion of long-term locked-in patients report a meaningful life and their demands for euthanasia are surprisingly infrequent. Both results confound the common notion that such a life could be "not worth living." I find this result astonishing, yet reassuring in the face of the many patients and families that I have encountered over the years. I find myself thinking, How can this be so? How can so many of these people be *happy*? It makes no sense.

As Laureys and his team write in their paper, perhaps the "happy" locked-in patients have recalibrated their needs and values. Like the Paralympian who finds victory in the face of physical adversity, it seems that these people have found new ways of experiencing life, new ways of achieving happiness.

This study questions whether any of us are in a position to judge what we might want to happen to us following a serious brain injury. Is it dangerous, then, to make an advanced directive? Imagine the nightmare of leaving a "do not resuscitate" order and being conscious as it was carried out against your (current) will.

Technology is advancing rapidly, and the day will come when we will be able to detect awareness, where it exists, at the bedside (or even at the roadside)—reliably, cheaply, and efficiently. We will be able to find those patients who are there, make contact with them, and assess their wishes. Whether we will be able to act on them is, however, an entirely different matter.

Abraham remained in hospital and eventually died from long-term complications arising from his stroke. Six months later, his wife seemed to be coping well with her loss as her church and family rallied around her.

Keith was allowed to die in 2013 according to his wife's wishes. His brothers and sisters were invited to the funeral.

At the time of writing, Hassan is alive and living in a hospital in Toronto.

CHAPTER TWELVE

ALFRED HITCHCOCK PRESENTS

≈

I have the perfect cure for a sore throat: cut it.

—Alfred Hitchcock

By 2012 we'd been asking people to imagine playing tennis in scanners for seven years. We'd established that a significant minority of patients who, like Carol, were assumed to be in a vegetative state could change their pattern of brain activity to indicate that they were actually conscious and aware. Almost 20 percent of them. We'd even asked a few highly recognizable figures—from Anderson Cooper to Ariel Sharon—to imagine playing tennis in the scanner. Some of our patients in the gray zone, such as Scott, had even been able to communicate with the outside world by simply *imagining playing tennis.* Imagining playing tennis had gone from being a quirky little idea dreamed up one summer in the garden of the Unit in Cambridge to a veritable cottage industry of research and media activity. It seemed to be the perfect solution to an enormously troubling clinical problem—finding people who were locked in the gray zone. Except it wasn't.

A pattern of data was beginning to emerge to suggest that we could do better. We'd seen several patients who hadn't been able to imagine playing tennis in the scanner—or at least, we hadn't been able to detect whether they were imagining playing tennis—yet they could do other things to show us that they were aware. We had no idea why. Martin Monti had developed a quite brilliant fMRI task back in Cambridge that showed that some of these patients could direct their attention to a face or a house when asked to do so—clear evidence that they could follow commands. This task relied on the fact that our brains have specialized regions for processing information about faces and information about places.

If you'll recall, face perception activates an area of the brain called the fusiform gyrus. Kate's fusiform gyrus had sprung to life when we showed her photos of faces in 1997. Another area of the brain known as the parahippocampal gyrus processes information about places. In 2006 Carol had activated that part of her brain when she imagined moving from room to room in her house.

Martin's experiment combined these two facts in a rather elegant way. He presented patients with a photo of an unfamiliar house superimposed on a photo of a stranger's face. In this superimposition, the patients focused on the features of the face (the eyes, the shape of the nose, and so on) or on the features of the house (the position of the front door and the number of windows).

This is much more easily accomplished than you might imagine. Despite being blended as one, the face and the house continue to exist as distinct images, rather than one image that is a combination of the two. You don't see a house with eyes, for example, or a face with windows. You see a complete face or you see a complete house, depending on the focus of your attention. If you focus on the windows, you will see the house, and the face will become virtually invisible. If you instead focus on the eyes, the house will become almost invisible. You can more or less re-create this effect for yourself by sitting in the front seat of a car and looking through the windshield at an outside object—say another car.

Although this blended scene (the windshield and the other car) forms a single image on your retina, your brain processes them as separate and distinct. You don't see a windshield with a car superimposed upon it, you see either the windshield or the car depending on where you choose to focus your attention.

What Martin showed with his elegant experiment is that if you ask healthy participants in the fMRI scanner to focus first on the face and then on the house, activity in their brains would switch from the fusiform gyrus to the parahippocampal gyrus, exactly at the point that they made the attentional switch. The amazing thing about this is that the stimulus (the blended image of the face and the house) had not changed one bit—all that had changed was what aspect of the image the participants were attending to. It was a measure of their ability to follow commands—just like imagining playing tennis or imagining moving from room to room in your house. Martin found that some of our patients could do this switching task on command but could not do the tennis task. We have no idea why, but my intuition was that switching between imagining playing tennis and imagining moving from room to room in your house was rather too *cognitively demanding* for some of our patients. It requires *too much effort*. Especially when you have to do it every thirty seconds for a long and boring five minutes in an fMRI scanner. We know for sure that brain injury in all its guises reduces the ability to do cognitively demanding tasks.

Even mild brain injury, which might not seriously affect your ability to perform many tasks, will almost certainly affect hard tasks more than easy ones. Again, this is because hard tasks, such as mental math, require more of our cognitive resources—more brainpower, for want of a better expression—than easy tasks, such as remembering a person's name. Think about what happens when you haven't had much sleep and you're trying to get through the next day. Easy (or well-practiced) tasks such as feeding the cat or even driving your car are simple to accomplish because they don't put too much demand on your depleted cognitive resources. But try filing your tax return or organizing a family vacation and you'll soon run into trouble. That's because these are more *cognitively demanding* than feeding the cat or driving your car, so they get hit hardest when your brain is functioning less than perfectly—such as when you haven't had enough sleep or after you have sustained a serious head injury.

Simply switching between looking at a face and looking at a house certainly feels less cognitively demanding than imagining playing a vigorous game of tennis for thirty seconds at a time. Perhaps imagining playing tennis was just too hard for some patients. They were slipping through our net, not because they were *not* conscious, but because the task we were asking them to do to show us that they *were* conscious was too hard for them.

Although Martin's task was easier, it had its own problems. Focusing on a face or a house requires you to have pretty good control over your eyes, and most of our patients just didn't have that. They couldn't control where they looked, let alone which aspect of a blended image they attended to. Clearly, we needed a different type of scanning task, a task that would catch all the conscious patients all the time, whether they had depleted cognitive resources or not.

≈

One of the postdocs who came with me from the UK to Canada was Albanian Lorina Naci. Back in Cambridge, she had wed my friend and colleague Rhodri Cusack. I was an official witnesses at their marriage. Rhodri was offered a faculty position in the Brain and Mind Institute at Western in 2011 and moved his lab there too. Lorina was therefore able to move at the same time. They have a son named Calin, who is a few months younger than my son, Jackson.

Since Rhodri's arrival at the Brain and Mind Institute, Lorina, Rhodri, and I had been trying to develop new and simpler ways of detecting consciousness. Our focus was on methods that are simpler for patients to perform so we could more or less automatically detect consciousness in an otherwise nonresponsive body rather than on methods that required patients to "report" that they are conscious.

Theoretically, this is an important distinction, and one that was becoming increasingly important for our research. Tests such as the tennis task do not *measure* consciousness in any direct way and do not tell us anything particularly important about consciousness itself, other than that it is present. Likewise, Martin's task with the superimposed faces and houses. These methods measure what philosophers like to call reportability—in this case, the ability to report that you are conscious. The problem is, it's perfectly possible that people exist who are conscious but are nevertheless not able to report that even by using their brain in an fMRI scanner, perhaps because they don't quite have the cognitive resources necessary to go that extra step. Just because they can't tell us that they are conscious does not mean that they're not. Philosophically, we were wrestling with the same issue we'd been trying to tackle for years: How do you measure consciousness in the absence of reportability? We'd traditionally been dealing with a lack of *physical* reportability, but perhaps *mental* reportability was just as problematic?

For Rhodri, this was important for his own line of research—using fMRI to try to map the development of consciousness in newborn babies. Both Jackson and Calin had more MRI scans before they were a year old than most adults will have in their lifetime. Babies are an excellent example of individuals who may be conscious—they certainly have some aspects of consciousness—yet they are not able to report that they are conscious because they do not yet possess the introspective abilities, or the language skills, to make such a report. Put simply, you can't ask a toddler to imagine playing tennis because most won't have a clue what tennis is, nor what you mean when you ask them to "imagine" something. To assess infant consciousness effectively you can't rely on *reportability*; you need a more direct readout of consciousness as it occurs in the brain.

By 2012 we were beginning to think we needed the same type of approach for patients who appeared to be vegetative. Rather than asking them to perform a task in the scanner, such as imagining playing tennis, we needed a more direct measure of consciousness; more direct and a whole lot simpler than the tasks we had hitherto deployed.

Our quest led us in a new and exciting direction. We began developing techniques to detect consciousness by showing patients Hollywood movies in the scanner. The idea came from a study that had been conducted almost ten years earlier by colleagues in Israel that had nothing to do with brain injury or disorders of consciousness. They had shown healthy participants movies in the scanner and noticed that as the plot unfolded, everybody's brain synchronized, with the same regions turning on and off at the same points in time. On the face of it, this makes perfect sense. When a gun goes off in a movie, our auditory cortex, the part of the brain that detects sounds, will activate; and because every person in a movie theater hears the gunshot at the same instant, all of their auditory cortices will simultaneously activate.

The same is true of many other events that are common in film. For example, when a face looms large on the screen, the fusiform "face area" will be activated in each and every person looking at that face. As the camera takes us through a scene, perhaps moving from street to street in a speeding car, our parahippocampal "place area" will fire in synchrony with that of the person sitting right beside us, as each of our brains maps and encodes each location that we pass through. Thus, during a movie, countless regions of the brain will switch on and off in unison across a group of people, reflecting their shared conscious experience of the events unfolding on the screen.

This remarkable phenomenon—that all of our brains synchronize when we watch the same movie—gave Lorina, Rhodri, and me an idea that would completely change for the next few years how we measured consciousness in the gray zone. If we scanned vegetative-state patients watching a movie and their brains synchronized to those of healthy participants watching the same movie, wouldn't that be reasonable evidence that the patients were having the same rich conscious experience? And if they were having the same rich conscious experience while watching a movie, wouldn't it be reasonable to conclude that they were having a similarly rich conscious experience of their own life? A movie is often simply a portrait of another life, particularly if its plot revolves around human relationships. When a movie engages you, it captures your consciousness; you are there, in the movie, in the moment, and the real world outside that little bubble of consciousness evaporates. Great movies capture our attention and take control of our conscious experience.

We suspected that we might have stumbled onto a more direct measure of consciousness that was a whole lot simpler than imagining playing tennis. All we needed to do was to show vegetative patients a movie and watch their brains with our fMRI scanner. If their brains followed the movie in the same way that the brains of healthy individuals did, then it would be a good indication that they were conscious.

Lorina set her mind to solving all the theoretical and practical problems until we had a workable experiment. The biggest problem was, what movie to choose? We tried several and some worked better than others. We had high hopes for *The Circus*, a 1928 Charlie Chaplin classic, which includes a hilarious scene in which Chaplin gets trapped in a cage with a lion. The participants enjoyed the movie, but unfortunately the synchronization between their brains was not as strong as we needed it to be. For our purposes, the best movies had a strong plot, a clear and evolving narrative, with distinct characters that had well-defined roles.

This made some good sense. If you are going to capture everyone's consciousness in the same way, then you want to force everyone's attention to move from place to place and person to person at the same time, and you want everyone to experience the twists in the plot at the same time. You want to keep everyone's brain maximally engaged and, as much as possible, *similarly engaged*. On top of that, a generous helping of cinematic tension seemed to help. This last element led us to Alfred Hitchcock, the Master of Suspense.

The brain loves movies by Alfred Hitchcock. More so, it turns out, than many other movies. That's probably because they are constructed to make us think, fear, anticipate, expect, and react. Hitchcock's movies are designed to give viewers a shared conscious experience driven, in large part, by the recruitment of similar brain processes, as each viewer observes the events unfolding and seeks to understand their relevance, leading to an ongoing involvement in the plot. Hitchcock's suspense arises through understanding the plots' twists and turns, rather than through a series of fast-moving bangs and flashes that are the central components of many more modern (and I would say inferior) films. Those bang-and-flash movies also drive the brain, but not to the same extent as the subtle changes in direction—and misdirection—that are Hitchcock's hallmarks.

We chose a short black-and-white Hitchcock movie, made

for TV in 1961, called *Bang! You're Dead.* The irony was not lost on me. The movie depicts a five-year-old boy who finds his uncle's revolver, partially loads it with bullets, and plays with it at home and in public, unaware of its power. The extremely engaging plot gradually unfolds, with the viewer becoming more and more convinced that the gun will be fired and kill someone.

Lorina scanned a group of healthy participants watching the movie and it worked like a dream—the activity that we saw was highly synchronized as each brain responded similarly to the plot's tense twists and turns. We had our movie! Now all we needed was a patient.

≈

In August 1997, when he was eighteen years old, Jeff Tremblay was assaulted outside a friend's house in Lloydminster, Alberta, a small city about two hours east of Edmonton. According to his father, Paul, an operations coordinator for Husky Energy, Jeff was an outgoing teenager with plenty of friends, a hard worker who was saving money but wasn't sure what he wanted to do after he had graduated from high school that spring.

The evening that changed the family's life forever started out at a nightclub. Jeff was getting friendly with the ex-girlfriend of a former bouncer for the club, who was there that night. Jeff and the girl left the club to go to a friend's house to watch a movie. The ex-bouncer followed them and "called Jeff out," Paul says. The ex-bouncer knocked Jeff to the ground and, as Jeff was getting up, kicked him in the chest. The kick caused a cardiac arrest and Jeff collapsed. He was taken to the local hospital in Lloydminster and then airlifted to Edmonton.

Paul, who was out of town for work, learned about the incident the morning after it happened and immediately flew to Edmonton.

He found his son comatose and on life support. The thinking at the time was that people in Jeff's condition did not recover, and if they did, they remained in a vegetative state. Some of Jeff's doctors urged Paul to consider pulling the plug.

Jeff emerged from his coma after three weeks and began to breathe on his own. His wake-sleep cycles returned. But he was nonresponsive, and he was diagnosed as vegetative.

Paul shuttled back and forth between Lloydminster and Edmonton. When Jeff first came out of the coma, "he looked glazed," said Paul. "There was no life in his eyes. No expression. Nothing."

Then one day Paul was sitting at the end of the bed in a chair, watching his son sleep. "I was doing a crossword puzzle. You're praying day to day that there will be changes. But there was nothing. I looked up, and Jeff opened his eyes and looked at me. And suddenly there was this great big *smile*! There was life in his eyes. It was amazing, as if between the time he fell asleep and the time he woke up, a wire had connected. He *recognized* me. I knew he was back. It was as though he had gone somewhere very, very far away and returned."

Still, the vegetative-state diagnosis remained. Jeff could not respond to command, and doctors saw no evidence of the connection Paul had clearly detected and strongly felt.

Jeff came back to Lloydminster, where he lived at the Dr. Cooke Extended Care Centre.

≈

In 2012, fifteen years after Jeff's assault, Paul was still researching brain injury, hoping desperately to come across something that would help him show that his son was still there. By then, Jeff was in his midthirties, physically healthy, but unable to speak or follow basic commands.

Paul came across a story online about the research in my lab and immediately fired off an e-mail: "I would very much like to have Jeff tested for his awareness. It would make both Jeff's brother and myself extremely happy knowing Jeff understands what we are saying to him. I believe it would make Jeff feel better as well. I am certain that Jeff understands what I say to him but I have no way of knowing for sure. I want to know if he is in pain, if he is happy or sad and if he knows how loved and missed he is. I would be willing to do whatever it takes to get this test to happen."

We agreed to evaluate Jeff, and Paul arranged for his son to be transported two thousand miles via commercial aircraft in July 2012 from Edmonton, Alberta, to Hamilton, Ontario, which is about eighty miles from London. An ambulance brought them to Parkwood Hospital, where Paul made Jeff comfortable before retiring to the Best Western Plus Lamplighter Inn, located right across the street.

Paul recalled the trip: "Jeff's reaction to the whole experience was amazing. When the stewardess explained the safety regulations, he turned his head and focused on her. I felt he was aware of everything, and I was amazed by how smoothly it went."

When Jeff was safely installed at Parkwood, my team assessed him. We asked him to look at a pen—nothing. We asked him to look at a mirror—still nothing. We asked him to stick out his tongue. No response. Curiously, he did show some evidence of "visual tracking": when a playing card was moved in front of his face he seemed, on occasion, to follow it with his eyes. Clinically, that placed Jeff in a "minimally conscious state." Nevertheless, my team found no evidence of awareness, nor any indications that Jeff was able to communicate.

But one thing we did learn set us all back on our heels: the weekly ritual that Paul had established for his son. Every weekend, for more than a decade, Paul had brought Jeff to the movies, wheeling him through downtown Lloydminster in his red-cushioned wheelchair to the May Cinema 6 multiplex. Incredible as it seemed, Paul was convinced that Jeff—who to us seemed minimally conscious *at best*—was absorbing everything on the big screen. According to Paul, Jeff generally preferred comedies and was a big *Seinfeld* fan. While part of me wondered if Paul might be deceiving himself about Jeff's awareness, another part found Jeff's supposed preference for *Seinfeld* interesting. *Seinfeld* has little broad physical comedy; the humor can be rather subtle and based on relationships that are established and evolve over time.

We sent another ambulance to Parkwood to pick up Jeff and Paul the following day and bring them to the scanning center. Paul, a good-looking tall man with a full head of dove-gray hair, stood beside the gurney as Jeff was wheeled in through the secure heavy door that separates the scanner from the rest of the world. Jeff had a lean face and closely cropped hair. He was alert, wide-awake, his head cocked to one side as he sat propped up on the gurney's pillows. I thought about how much love and commitment it must have taken for Paul to make this journey with his son and hoped that we might be able to send them both home with some good news. I told Jeff about the fMRI scan and the movie that he was about to see. It was a strange, almost cinematic moment. That we were about to try out our new Hitchcock task on this particular patient—a seasoned moviegoer by any measure—felt like the sort of coincidence that could only happen in the movies!

As Jeff slid into the scanner, I couldn't help but wonder whether Alfred Hitchcock would be the one to finally give Paul what he needed—evidence that his son Jeff was conscious and aware. What an odd irony that would be. All those weekends, all those movies. What if Jeff had experienced them all, just as you and I, while those around him remained blissfully unaware that he was aware of anything at all?

I stepped outside to the waiting room, where Paul was patiently waiting. "We're just showing Jeff an Alfred Hitchcock movie," I told him, "to see whether we can activate his brain."

Back in the fMRI room, *Bang! You're Dead* was playing on the screen above Jeff's head. We knew the screen would be visible to him via a mirror mounted in front of his eyes, but we couldn't be sure he was watching. When it was over, we pulled Jeff out of the scanner and sent him back to Parkwood for the night.

≈

It took a few days to analyze the data. The procedure was more complicated than the one we'd been using for the tennis task, and Lorina was still trying to iron out the kinks. There wasn't a road map for this kind of thing. How do you examine the brain of a person watching a movie and determine whether the person is *consciously experiencing* it? We didn't know because no one had done it before. We'd run the analysis in controls, but we *knew* they were conscious—this required something more. We had to develop the methods as we went along. When we had the results, I was stunned. Although Jeff's brain activity was a little reduced compared to that of our healthy controls, as he watched the movie, all the appropriate brain areas activated *at the right time*. In response to sounds, Jeff's auditory cortex sprang to life. When the camera angle changed or the young boy ran across the screen, Jeff's visual cortex activated. But most important, at all the critical twists and turns in the plot—those places where a clear understanding of the story unfolding on-screen is essential—Jeff's frontal and parietal lobes responded exactly like those of a person who was conscious and aware. Jeff was watching the movie! More than that, Jeff was *experiencing* the movie! We had used an Alfred Hitchcock movie to show that Jeff, who was presumed to have been vegetative for fifteen years, was conscious and experiencing the movie just as you or I would. All those weekends, all those movies, all Paul's efforts had not been in vain. And we had deduced that based solely on Jeff's brain responses.

≈

How did we know that Jeff was really conscious? As ever in science, the devil was in the details, and in this case the details came courtesy of Mr. Hitchcock. *Bang! You're Dead* engages the parts of the brain that we know are involved in everyday conscious experiences. Lorina's studies in healthy participants had already shown us that. A movie with lots of loud bells and whistles will undoubtedly stimulate the auditory cortex, although that does not mean that a patient is conscious, as we had seen from our scans of Debbie and then Kevin. Similarly, a movie with lots of changes in light and dark, lots of movement and changes of scene, will activate the brain's visual cortex—but again, this would likely reflect an automatic brain response and bear little relevance to whether the patient was consciously *experiencing* those changes.

Bang! You're Dead was much subtler than that, subtlety that we could turn to our advantage. Specific elements are intrinsic to the plot. The gun and its potential to shoot people. The circumstances of the main characters (they are capable of shooting or being shot). And what psychologists call theory of mind—the ability to attribute mental states to other beings and to understand that they might have beliefs, desires, intentions, and perspectives that are different from our own. To fully appreciate *Bang! You're Dead*, theory of mind is essential because you have to realize that, although you (the viewer) know that the gun is real, the young boy thinks that it's a toy. That is why the situation is so suspenseful—the boy loves playing shoot 'em up with his little cowboy friends, but this time he doesn't know it's for real. But you do!

Many areas of the brain are known to be responsible for giving us theory of mind, but one region that seems to be essential is a part of the frontal lobe toward the front and in the center of the two cerebral hemispheres. In 1985, my Cambridge colleague Simon Baron-Cohen and his colleagues were the first to suggest that children with autism lack theory of mind. Many of their problems appear to stem from a lack of understanding what those around them are thinking. Indeed, whether normally developing children younger than three or four years old have a theory of mind is hotly debated, as is whether nonhuman species do.

In addition to theory of mind, watching *Bang! You're Dead* invokes a whole variety of other complex cognitive processes that are relevant to, and indicative of, consciousness. For example, you have to draw on your long-term memory to understand what the boy is holding (a loaded gun) and what it is used for (to kill people). If a person who had never seen or heard of guns before were to watch the movie, they would not be afraid because they would have no sense that what the boy was holding was dangerous. The boy might as well be waving about a banana!

Our elaborate knowledge about guns is what scares us about a child with a gun. Guns kill people and start wars. We also have an elaborate theory of mind about children: they don't understand guns, they don't understand that they kill people and start wars. This knowledge is fundamental to our sense of suspense. An unloaded gun in the hands of a child is not scary. A gun, loaded or unloaded, is less scary in the hands of an adult (particularly a responsible one) than in the hands of a child. A gun, loaded or unloaded, is no more or less scary to a monkey than a banana (unless that monkey has witnessed, and learned, from seeing hunters kill other monkeys with guns) because monkeys don't have this dense background knowledge that generates our conscious sense of the world, in this case our suspense from seeing an innocent child with a loaded weapon. It's fascinating that our consciousness—or, perhaps, a better way to put it would be that our conscious sense of the world around us—is not generated by who or what we are, but by our *experiences.*

≈

Jeff's remarkable response in the scanner to *Bang! You're Dead* was a theoretical milestone for us. We had shown for the first time that the brain activity produced by similar conscious experiences in different individuals could be used to infer conscious awareness in physically nonresponsive patients without any need for self-report. All Jeff had to do for us was lie in a scanner and watch a movie. To be clear, we weren't reading the precise details of his thoughts, but showing that his thoughts, whatever they were, were highly similar to a perfectly healthy person's thoughts while watching the same movie.

When we published Jeff's story and our new approach to measuring consciousness in the prestigious journal the *Proceedings of the National Academy of Sciences* in 2014, another wave of intense media attention followed. Lorina appeared on several TV news shows and spoke to radio stations and newspapers around the world. The response was overwhelmingly positive. It seemed that in the years since we'd shown for the first time that neuroimaging could be used to detect hidden consciousness in some patients who are assumed to be vegetative, the media, and the scientific community, had become accustomed to the idea. Detractors were few, if any.

Our findings were particularly important to Jason, Jeff's brother: "I talk to him with more passion now. I still have my wonders about what gets through to him and what doesn't."

Jason tells his little brother "to keep fighting. To not give up. I don't know if that's selfish of me. It's hard to lose someone and not really lose them. I want him to know how much he means to me. This is the new version of Jeff. This is who he is."

Now Jason knows that Jeff understands what he was trying to tell him. "When you're eighteen and twenty-one, you don't say things like 'I love you,'" Jason said. "Your tests reaffirmed all those talks I had with him in private. To know that he's heard me—it does feel good."

CHAPTER THIRTEEN

BACK FROM THE DEAD

≈

> Everything dies, baby, that's a fact,
> But maybe everything that dies someday comes back.
>
> —Bruce Springsteen

On July 19, 2013, Juan spent the evening with friends, returning home around midnight. He made himself a snack, said goodnight to his parents, and turned in. Everything seemed normal. But at 6:30 a.m. the following morning, things were far from normal. Margarita awoke to the sound of her nineteen-year-old son choking to death in his bedroom, just a few yards away. She rushed into his room and found him unresponsive, lying facedown in his own vomit.

Juan was rushed to his local emergency room in a hospital south of Toronto. A CT scan showed extensive damage to the white matter in his brain, including the frontal and parietal lobes, regions critical for working memory, attention, and other high-level cognitive functions. The occipital lobe was also affected—the very back part of the brain that is crucial for vision. A structure deep within the brain known as the globus pallidus was also badly damaged. The globus pallidus plays a vital role in voluntary movement, and disruption of its normal functioning is one factor that causes the symptoms of Parkinson's disease.

This kind of brain damage, widespread and diffuse with no clear borders between healthy and damaged tissue, is common when the brain has been starved of oxygen. When the oxygen dries up, the brain starts shutting down little by little, piece by piece, until not even enough functional tissue is left to keep our most primitive bodily functions, such as breathing, going. Juan wasn't quite there, but he was close. On admission, he had a Glasgow Coma Scale score of 3 out of a possible 15. You can't score lower than a 3, not without being dead.

Two months later, Juan remained totally unresponsive to any form of external stimulation and was declared to be in a vegetative state. He was fed and hydrated through a tube. His parents, who had remained at his bedside since day one, brought Juan from his local hospital to see us. They were hoping that we could tell them more about his condition, perhaps even make some predictions about his future.

To my team, Juan appeared no different from most of the patients we see: awake, but seemingly unaware, completely nonresponsive. We took him for fMRI scans, hoping they would tell us more about the state of his brain and the likelihood of some recovery. We asked him to imagine playing tennis. Nothing. We asked him to imagine walking around the rooms of his home. Again, nothing.

Lorina tried him on the Hitchcock task. Would Juan's brain respond to the twists and turns of *Bang! You're Dead*? The results were mixed. Juan's auditory cortex was clearly responding to the sound track of the movie; but curiously, his occipital lobe, the part of the brain responsible for vision, showed little response. Perhaps Juan's extensive brain damage, including to the occipital (or visual) cortex, had left him blind? There was no way to know. But if Juan couldn't see the movie, then he couldn't follow the plot, making it more or less inevitable that we would see no activity in frontal and parietal regions of his brain—the activity that we needed to see to determine whether he was conscious. Two days later, we put Juan back into the scanner and repeated the whole procedure. Every patient deserves a second chance. We threw everything we had at him, but again, we got nothing back.

After four days, Juan returned home with his parents, as much a mystery to us as he had been when he arrived.

≈

Seven months later, Laura Gonzalez-Lara, my research coordinator, called Margarita to follow up on Juan's progress. We do this with all our patients, in part because some do improve over time and we try to monitor that as closely as we can, and in part because it is a way for us to stay in touch with our patients' families. I've never been comfortable sending a family away following an assessment with a simple "Thank you very much, there's nothing more we can do." Often that may well be the case, but it just doesn't feel right to offer nothing: no follow-up, no further investigations, no hope.

"How has Juan been?" Laura asked.

"Why don't you ask him?" Margarita replied.

Against all the odds, Juan was talking, brushing his teeth, eating, and walking.

When Laura reported this to me, I fell off my chair. I couldn't believe it! "You mean he's recovered? He's come back from the dead!" I exclaimed. When I get excited, I am famously prone to hyperbole.

"Apparently so," replied Laura, characteristically understated.

I had never seen—or heard of—*anything* even remotely like Juan's recovery. Occasionally patients improve from a vegetative to a minimally conscious state, transiting from "nonresponsive" to "partially responsive, some of the time." This was something else. Like my first patient, Kate, Juan was talking again. But unlike Kate, he was also *walking*.

Juan's unprecedented improvement made me wonder whether he was really in a vegetative state when he was scanned. Had he really come back from the gray zone, or was it possible that he was never there at all? Perhaps he'd just had some sort of temporary physical paralysis—an inability to move his limbs that gave the impression that he was vegetative when, in fact, he was just nonresponsive. I checked his medical records—we had obtained copies of all of his tests and scans from his referring physician. The circumstances of his case were clearly described by several neurologists and therapists who had examined him during his illness. Everybody agreed that Juan had sustained severe brain damage that had left him in a vegetative state. And the CT scan revealed just how extensive that damage had been.

I called an emergency lab meeting. Everyone working with our brain-injured patients, whether they had seen Juan or not, gathered around the large cluster of tables in the small seminar room at the Brain and Mind Institute at Western—at least a dozen colleagues, students, and postdocs. I wanted to get as many opinions as possible. Clearly we had to get Juan back to London to reassess him as soon as possible. If we dawdled, he might move on with his life and have no interest in helping us answer the questions we were dying to ask. Worse still, he could relapse and go back into the state he'd been in when we'd first assessed him seven months earlier.

I knew exactly what I wanted to know. Did he remember anything from the period when he'd come to London to be scanned the previous year? This wasn't just idle curiosity. In all the years we'd been seeing patients who turned out to be more conscious than they appeared to be clinically, I'd never encountered one who could report their experience in the scanner at a later date. What was it like to be conscious when the people all around you think that you're in a vegetative state?

Had Juan tried to move? To talk? To signal in some way that he was still there? I wanted to know how it *felt* to be where he was with all the clinical apparatus and diagnostic tools we set in motion around cases such as his. Even more important, what could be more convincing evidence of consciousness than a firsthand personal report? If Juan could describe the novel and unusual experience of lying inside an fMRI scanner, then we would know that he must have been conscious when it occurred. How else could he know what that experience is like? In Juan's case, this was important because his scanning data had been so inconclusive. We had no evidence from the scans to suggest that he had been conscious—what better solution to that than to get him to tell us himself?

We set about devising a series of tests for Juan to see whether he could remember anything of his experiences with us. This was not quite as scientifically straightforward as it might seem because we had to reconstruct his entire visit seven months earlier, just to establish what we should be asking him. Imagine that you were introduced to a stranger and you had to work out whether the two of you were present at the same event—perhaps a party—seven months earlier. How would you do it? Would you begin by asking if he remembered who else was there? Perhaps you'd show him a photo of the apartment where the party took place?

The problem with this approach is what you do if the results are negative. Just because he didn't recognize someone who attended a party, or the apartment where the party was held, does not mean he was not there. Perhaps he wasn't observant or has a poor memory for such things. I barely remember whether I went to any parties seven months ago, let alone who was there, or where they might have been held. And even if I did remember going out to a party seven months ago, whether this person or that person was at that particular party or a different one is beyond me.

This is an odd kind of memory problem, remembering who was present on a specific occasion and what the environment looked like. If we only had one thing to remember, one face in one place on one occasion, it would be easy. The problem is that over a year most of us attend several parties with different casts of characters, some of which may be in novel and unique locations, while many are not. All this causes what psychologists call *interference*: blips in our memory about who was where and when. Our recall becomes slightly confused over time.

Fortunately, in Juan's case, we had a number of factors on our side. For most of us, being in an fMRI scanner does not happen as often as we go to parties (although for some notable exceptions to this rule, check out almost any member of my lab). For Juan it was certainly a once-in-a-lifetime experience. Similarly, the other tests that we performed that week—the neurological examinations and the electroencephalograpy (EEG) assessments—were all likely to have been unique events for him that would not be subject to interference from other similar occasions. Pretty much anyone he had seen that week and any place that he'd been would be a unique experience that we could use to probe his memory. We still had the problem that if he didn't remember anything, it didn't necessarily mean that he had been unconscious at that time; but if he did at least remember being in the scanner, meeting my students, and being asked to watch the Hitchcock movie, we'd have good evidence that he had, indeed, been conscious.

We made a list of all the places we'd taken him in London—the hospital, the ambulance, the scanning suite at the Robarts Research Institute—and a list of the people who had assessed him:

Laura, my research coordinator; Steve, a graduate student working on his master's thesis; and Damian Cruse, one of my postdocs who ran the EEG lab. We found pictures of those places and pictures of those faces. Then we assembled a matched series of "control" places and faces. Pictures of places Juan hadn't been, such as the experimental testing rooms in the Brain and Mind Institute, and graduate students who were working on other projects in the lab at the time and had not been to see Juan during his visit.

We had to get this right because we had just one shot. We had a limited number of people and places to choose from, and once we'd shown Juan pictures of them, we'd never again be sure whether he was recalling them from his first visit as a vegetative patient or whether he was simply recalling the pictures that we'd subsequently shown him while trying to test his memory.

≈

Juan and his parents came to London, and he was admitted to Parkwood Hospital. As he sat in his wheelchair waiting for the memory test, Juan remained strangely serious, almost gloomy. As I look back, it seems odd that someone who had turned his life around so dramatically would not be ecstatic, thankful for every day that he had clawed back from the void. But Juan was quiet and detached. Perhaps it was all part of his recovery. Perhaps only some parts of Juan had come back—maybe some part of his personality had been left behind. Or perhaps he just needed more time.

We were all on tenterhooks. The testing room's atmosphere was electric. Steve and Damian administered the memory test that we'd hastily, but carefully, put together just for Juan. His answers were astonishing. Yes, he remembered being scanned—going into a dark tube and being afraid. He remembered the Hitchcock movie. He described Laura's facial characteristics in exquisite detail and clearly remembered Steve, who had tested him with the EEG. In that first week we'd tried out some of our new EEG techniques on Juan, as well as two fMRI scans and a series of behavioral evaluations, in the hope that one of these approaches would give us a positive result.

Juan recalled Steve: "He put electrodes on my head and had a deep voice." Steve does indeed have a deep voice and "he put electrodes on my head" is as good a lay description of EEG as I have ever heard. Juan remembered everything about his first visit, down to the tiniest detail.

I can't emphasize enough how extraordinary that was. Over the years, we've seen many patients who've sailed through all the standard clinical tests and then been placed in the vegetative category, only to find that they can imagine playing tennis or produce other responses in the scanner that tell us that they are, in fact, conscious. But to recover and tell us all about their experiences in the scanner? That had never before happened. Not even close.

We finally had absolutely unassailable evidence that a patient could appear to be entirely vegetative, yet remain absolutely conscious, experiencing life down to the very last detail without any of us even knowing it. Think about it. How else could Juan describe the inside of an fMRI scanner unless he'd been there and awake when we pushed him into it? How else could he have known which movie we used to activate his auditory cortex unless he'd experienced it? How would he know Steve, someone Juan had never encountered before the day he came to London and had not again met since the start of his remarkable recovery? The only explanation was that Juan had defied medical opinion and continued to monitor and *remember* the world around him for many months, all the while appearing to be in a vegetative state. What was perhaps *most* remarkable about this feat was just how good Juan's memory for that period was. His brain had been starved of oxygen and sustained massive damage. How was that possible?

The more I thought about Juan, the more I realized how little we still understood about consciousness and its many faces. We'd thrown everything we had at Juan, every type of brain scan, every newfangled technique we had at our disposal; yet we had failed to spot consciousness where consciousness clearly existed, in spades. Weirder still, this unseen essence of Juan—this part of him that was in there, *experiencing* the scans just as you and I would—had fought its way out of the gray zone. This haunting reminder of the resiliency of consciousness forced me to reflect anew on the nature of *being*, the meaning of what it means to be alive, and whether *anyone* can be said to be irretrievably lost. Maureen's scans had shown nothing. But so had Juan's. Might there *still* be some hope for Maureen and people like her?

≈

Many things about Juan remained a mystery. Why, if he was conscious and aware throughout his first trip to London, were we not able to pick this up with our fMRI scans? Why couldn't he imagine playing tennis or imagine moving around his home? Why did the Hitchcock movie only activate his auditory cortex and not the frontal and parietal lobes, which would have clearly indicated to us that he was in there, experiencing the twists and turns in the plot just as you or I would? On two separate days we'd scanned him, and on both occasions we'd drawn a blank. Again, negative findings in patients such as Juan are hard to interpret. We knew that he hadn't fallen asleep because on the computer monitor that was connected to the tiny camera inside the scanner we could see that his eyes were open. Besides, if he had been asleep, how would he have been able to recall the details of the scanning session in such exquisite detail? Perhaps, due to the particular nature of his brain damage, Juan was aware, yet somehow unable to generate responses at the appropriate time. Or perhaps his awareness fluctuated in and out, sometimes present—just enough to keep track of what was going on—and sometimes not. Perhaps he just didn't want to respond? We just didn't know. What we did know was that he had been conscious enough to experience, remember, and report almost everything that happened that day, regardless of what his brain did in the scanner.

≈

A little over a year after Juan's second visit to London, and his remarkable performance in our memory tests, I drove to his home to see how he was doing. Laura had been in regular touch with Margarita, so I knew he'd continued to make good progress, but I wanted to see him for myself and also ask some him additional questions that had been nagging at me.

I pulled into Juan's street; comfortable two-story homes were packed together in a planned suburban Toronto community. Margarita, a friendly dark-haired woman, led me inside. The house was adapted with ramps for Juan's wheelchair.

"He'll be a few minutes late," Margarita said. "He usually takes the bus to school. Today his dad is picking him up."

Juan takes the bus? On his own? To school? I found myself again in a kind of altered state—I couldn't quite believe what I was hearing. I knew that Juan had continued to recover, but this was well beyond my expectations.

I hope I did not seem too incredulous as Margarita and I chatted. "We were at a very dark point in our lives when we came to you," Margarita said. "You gave us hope. The doctors said his brain was done. Zero chance of recovery and no options. And then the person who managed the ICU mentioned you."

The front door opened, and Juan wheeled himself into the room. My astonishment—and curiosity—deepened. Juan was intense-looking, with trimmed dark hair and dark eyes, and his personality now broke through in a way that had been entirely absent when he'd come to London a year earlier.

"What do you want to talk to me about?" he asked.

I suggested he tell me about his experiences in the hospital soon after his accident, before he was referred to us for scanning.

"I felt like I was trapped. But I wasn't terrified or despairing. I knew that I was going to get through eventually." The words were emotional—some part of Juan, *the feeling part*, had returned.

"Presumably you were trying to move and speak?"

"I was trying to speak constantly."

"Were you in pain?"

"No. It was like being inside my body but not being able to control it."

"I'd touch his feet with ice," said Margarita. "Bring grains of coffee for him to smell. I was making a case to send him to the recovery center. I put together one hundred and twenty hyperbaric-chamber sessions all on my own."

Many families of patients who are declared vegetative resort to their own therapies, such as hyperbaric oxygen therapy, to which Margarita referred. In hyperbaric oxygen therapy one breathes pure oxygen in a pressurized room or chamber. This treatment is well established for decompression sickness, which can occur when scuba divers come up to the surface too quickly. In a hyperbaric oxygen therapy chamber, the air pressure is increased to three times that of normal air pressure, which allows the lungs to take in more oxygen than would be possible by breathing pure oxygen at normal air pressure—in short, it increases the amount of oxygen the blood can carry. Some evidence suggests it can be useful for treating serious infections.

Margarita and her family turned to it because no conventional treatments for Juan's condition existed.

"The hospital didn't know what to do," she said. "They kept piling on meds. In three months, seven cycles of antibiotics. His immune system was shutting down. He'd have a high fever for four or five days. The oxygen therapies helped his immune system become stronger. [I hired a] dietitian experienced with brain injury who was very specific with supplements. We did it ourselves. Juan is not a miracle but a lot of hard work."

The conversation moved back to Juan's memories and experiences.

"What do you remember about the first time we scanned you?" I asked him.

"I was afraid." Again Juan's words were imbued with feeling. I started to wonder whether Juan had returned from the gray zone in parts, bit by bit. When he'd come to London for his memory test a year earlier, some parts of him were definitely there—his body, his memory, his physical being. But some parts were definitely missing, and only now was it clear what they were. Juan the person had returned; Juan the personality. The essence of Juan was finally back from the gray zone, perhaps not completely, but enough to know that he was going to make it eventually. All of him.

Thousands of people, both patients and healthy volunteers, have been through our scanners. Although occasionally someone gets anxious, it's rare.

"Why were you afraid?"

"I didn't know what was going on."

I had to ask the next question: "Would you say that when we put you into the scanner that first time, we didn't tell you enough about what was happening?"

He looked directly at me. *"Definitely."*

I was horrified. Although we go to great lengths to tell our patients, whether they appear to be vegetative or not, what the scanning session involves, sometimes I guess we're not thorough enough.

It was worse. Juan continued, "I was so scared that I cried."

We routinely film our patients' faces through a tiny camera mounted in the bore of the magnet, and my team monitors patients closely. No notes suggested that Juan had cried during the scan.

"Did you cry tears?"

"I couldn't produce tears. But I still cried."

I shall always remember this heartbreaking moment when I prepare a patient—or anyone else—for the scanner. I probed more deeply: "Do you feel that you remember everything from that first visit?"

"Yes, everything."

I had little doubt that Juan was cognitively back to his former self. His answers were short—mostly single words—but they were efficient and complete. He was giving me just enough information to answer the questions, but never more than I requested. Occasionally, he'd let something slip. Some little nugget of information that would tell me that his worldview—his perspective on life and all that had happened to him—was entirely normal for someone in his position.

Over the next hour or so, Juan told me and showed me many incredible things. He pulled himself up from his wheelchair and shuffled, one step at a time, along the path created by a set of parallel bars that his parents had set up in a room off the kitchen.

I noticed that his left foot wasn't moving as smoothly as his right. "What does it feel like when you try to move your left foot?"

"Like I'm pulling it through."

"You mean it doesn't do what you want it to do?"

"That's exactly it."

"What about your right leg?"

"My right leg does what I want."

Juan painstakingly shuffled from one end of the parallel bars to the other and back again, slowly turned himself around, and dropped down into his wheelchair.

"Fantastic, Juan!" I said, then felt immediately foolish. My superlatives paled beside his achievements.

Prior to his injury, Juan had been a budding DJ. He was back at the mixing deck. He played us some of his tunes, slowly but surely moving the computer mouse to push notes in and out of the mix. His fine motor skills were fully restored, albeit a bit slow.

I asked him if he noticed cognitive deficits.

"Thinking. I'm slower than the other kids. But I get there."

Cognitive slowing (bradyphrenia) is common after brain injury and also in some neurodegenerative conditions such as Parkinson's disease, but I had never had a patient with a brain injury *tell me about it* before.

In Parkinson's disease, it occurs as part and parcel of the patients' main symptoms. Parkinson's patients move slowly, but they also think slowly, even after you've taken their slowed movement into account. Back in my PhD days we'd shown that when you give Parkinson's patients a simple problem-solving task, they take far longer than healthy elderly people to find the solution, although they do get there in the end. No one knows quite why this is—possibly the lack of dopamine in their brains that causes slowed movement also causes slowed thinking. As though every aspect of life is going along a little more slowly than before: there's still gas in the tank, but the brake is permanently on.

Juan didn't have Parkinson's disease, but in some ways his symptoms were similar. Perhaps the damage to his globus pallidus was the reason for the similarity. Juan's description "My left leg doesn't do what I want it to do" reminded me of the comments made by some Parkinson's patients. As though the leg no longer quite belonged to the patient. As though it has a life of its own.

I'd also heard something similar much more recently. Kate, the first brain-injury patient that we'd scanned in 1997, also described a sort of dissociation—or disconnection—between "her" the person and her brain when I saw her again in 2016. "My brain doesn't like me anymore," she had said. "It doesn't do what I want it to do."

Juan was also experiencing a dissociation, but in his case it was between him (Juan, the person) and part of him (Juan, the body). He didn't feel in control of his leg anymore. Despite his extraordinary recovery, Juan still felt that some part of him was somewhere else, outside his sphere of control, trapped in the gray zone.

≈

Juan was not the first person to have made a seemingly miraculous recovery, emerging from the gray zone and reentering the world. Jan Grzebski, a sixty-five-year-old Polish railway worker, made headlines when he "woke up" in 2007 after nineteen years in a coma, which he had entered as the result of a brain tumor. His world had changed beyond recognition. He remembered shops during the Communist regime that only had "tea and vinegar . . . meat was rationed and huge petrol queues were everywhere. Now I see people on the streets with cell phones, and there are so many goods in the shops it makes my head spin," he said on Polish television. He had also gained eleven grandchildren while in the gray zone.

Grzebski's case was a real-life rendition of *Good Bye, Lenin*, a German film that was an international hit. His remarkable story was reported around the world. The Fox News headline read, "Living Corpse Wakes."

Grzebski credited his wife, Gertruda, with his awakening. She would not give up on him, although doctors said he would never recover and gave him only two or three years to live. She moved him *every hour for nineteen years* to keep him from getting bedsores.

What an extraordinary act of love.

The tumor that had put him into a coma killed him in 2008, only a year after his "awakening."

In another well-documented case, Terry Wallis, an Arkansas man, suffered an acute brain injury when his truck skidded off a bridge in 1984. He was comatose after the accident and then minimally conscious. The prognosis was grim: doctors said he would never recover. Yet, mysteriously, in 2003 he went through a remarkable three-day period, an arc of "awakening," in which he gradually emerged from the gray zone. He thought it was still 1984 and that he was twenty! Nineteen years had passed in the blink of an eye. Where had "he" been all this time? What was going on in his brain?

Wallis's body had aged. The body continues to age in the gray zone, sometimes in an accelerated way from atrophying muscles. Wallis remained physically disabled and his short-term memory was shot, although he clearly recalled his life before the accident. As is the case with Juan, we have no idea what prompted his awakening. Or why he wasn't able to retain new information or experiences.

≈

Juan has given us an entirely new perspective on the gray zone. His recovery is singular—zero to hero, just like that. It doesn't get much worse than 3 out of 15 on the Glasgow Coma Scale, yet when I last saw him, he was mixing tunes like a professional DJ.

Margarita emphasized that the family's proactive, positive attitude had contributed to Juan's recovery. She had left her job for six months to focus on him and the extra therapies he received. They fund-raised through a website and collected $45,000.

It's hard to escape the thought that anyone could achieve the same miraculous result with enough willpower, love, and family support, enough money, enough luck perhaps. But I don't think so. Every brain is different, and every brain *injury* is different. The gray zone is an unpredictable place, mysterious and complex. We have learned a tremendous amount about it over the past twenty years and about the tenuous, fragile nature of consciousness, yet we still know so little about how and why some people recover and some don't. And even for those who do, the word *recovery* does not mean the same thing.

For the lucky few, recovery is like Juan. Back to college, riding the bus, hanging with friends. For others, recovery looks more like Kate, definitely back from the gray zone, reflecting on the hand that she's been dealt, coming to terms with what she's lost, little by little, day by day. But for most, the hard truth is that they gain a few extra points on the Coma Recovery Scale, a little bit more responsiveness. They move a few steps up the ladder, out of the abyss.

Several years ago, I stopped using the word *recovery* when I spoke to journalists. Not because I don't think anyone ever "recovers," but because the term is so strongly loaded for those of us who are relatively healthy. The word simply fails to reflect the expectations and achievements of those who are trying to "recover."

I "recovered" from cancer in 1981. I have a few residual health issues, but I'm essentially healthy and live a normal life. Recovery after serious brain injury is another matter. Few of the patients that I have seen return to anything resembling a "normal" life. Indeed, most don't recover at all. Juan, the best "recovery" story that I can tell after twenty years in this field, is the rare, rare exception that tells us that there is always some hope, however small. Juan has come almost all the way back from the gray zone, yet his experience there will have undoubtedly endowed him with a perspective and qualities that he didn't have before. Juan has seen things that most of us will never see in our lifetimes. Nor ever should.

Any sort of brain injury will likely have long-lasting pervasive effects. It's not the same for any other organ of the body. We can replace kidneys, lungs, hearts, and livers and essentially we are still ourselves—a little wobbly for a while perhaps, but the same person. Many of us return to live full and complete lives. Perhaps the same lives we would have lived had we not fallen ill, notwithstanding the emotional scars we inevitably carry when our lives have been threatened.

But serious brain injury is fundamentally different. It changes us, it alters our ability to move, react, interact, and respond. And recovery is far harder, if it occurs at all. We can't transplant brains (at least not yet), but even if we could, it wouldn't help us to recover in the way that transplanting a heart or a kidney helps us to recover. Because after a brain transplant, "we" would not recover; "we" would be someone else. We might look the same, but with someone else's brain in our heads we would be an entirely different person. Conversely, transplant your brain into another body and you would still be you—not that other person. You'd look different, and it's tantalizing to think that you might even feel different in ways both subtle and apparent. But you would be essentially the same person living in another body. The same thoughts, the same memories, the same personality. Your sense of being, the cascade of thoughts, feelings, and emotions that comprise our conscious experience of the world, would be largely identical. Like a perfect disguise, the appearance is different, but underneath the person is unchanged.

Kate told me that although her capacities have diminished, she is at her core the same person she was, deserving of the same love, attention, and respect that healthy people expect. Juan too, I'm sure, feels he is the same person, altered perhaps in ways that are beyond the measurable diminishment of physical and cognitive functions that are so hard to define. It amazes me that who we are, our very being, the very stuff that makes me, me and you, you, is phenomenally resistant to alteration, even by catastrophic brain damage.

There's no escaping it: we are our brains.

CHAPTER FOURTEEN

TAKE ME HOME

≈

I've seen the nations rise and fall
I've heard their stories, heard them all
But love's the only engine of survival

—Leonard Cohen

Juan's return from the gray zone was a sobering reminder that consciousness has always been one step ahead of us. With Alfred Hitchcock, we thought we had it, the perfect measure, an infallible tool for tracking down consciousness in its deepest, darkest elemental lair. But it had escaped us again, slipping right through our fingers. It was there, in Juan's experience, in its most elaborate of forms, yet we had failed to see it. fMRI is a tremendously powerful tool, and we were constantly refining what we could do with it. Increasing computer power had enabled us to ask questions of patients such as Scott and Jeff, moving us ever closer to the moment that we would be able to engage in a real-time two-way conversation with their inner selves. At the same time, our explorations into the gray zone were helping us to unravel the building blocks of consciousness—how brain processes like memory, attention, and reasoning relate to unitary concepts like "intelligence" and how they emerge from that three-pound lump of gray and white matter inside our heads (to see how we solved some of these questions, please visit www.cambridgebrain sciences.com). All around the world, we and others were using this extraordinary technology to map the architecture of our thoughts and feelings, identifying the crucial links between the way our brains function and how we experience our conscious world, how we develop a sense of identity, and how it is shaped by a lifetime of experience. Our adventures with the Master of Suspense had shown that our consciousness is tightly coupled to the consciousness of others experiencing the exact same event, to what we think others are thinking and feeling, to our *theory of mind*.

But fMRI was expensive to use, and moving patients to the scanner was difficult, limiting its potential to help people desperate to communicate with those near and dear to them who were marooned in the gray zone. A big part of the future of what we were doing clearly hinged on streamlining this cumbersome and expensive technology, making it portable and user-friendly, taking it out of the hands of scientists such as me and medical professionals and putting it into the hands of those who were so deeply invested in reclaiming the people who had been taken from them. And few people were as deeply invested as Winifred.

≈

One night in May 2010 at around 3:30 a.m., Winifred was suddenly woken by what she thought was her husband, Leonard, snoring in bed beside her. She must have intuitively known something was wrong. "He never woke me with his snoring," she said. "The joke was that the world could come apart and I would keep sleeping."

That night the family's world was, indeed, splintering. Somehow Winifred knew her husband was in trouble. She tried to rouse him, thinking he was having a nightmare. When she couldn't, she called out to her son and daughter, sleeping in nearby rooms. Her son dialed 911. Winifred and her children were told to move Leonard down from the bed so he was lying flat on the floor. This was no easy feat. Leonard was a large man who had been a sailor in his youth in Bombay and had worked in the shipyards of Dubai.

The ambulance arrived ten to fifteen minutes later by Winifred's calculation. "I've thought about how long it took for the ambulance to come, over and over and over in my head," she said. Leonard had stopped breathing. The medics quickly determined that he was in the throes of a cardiac arrest, administered CPR, and got his heart beating again, but he was slipping away fast. They rushed him to Brantford General, the local hospital, where he was placed in a medically induced coma to reduce the chances of further damage to his brain. After injury, the metabolism of the brain has often been significantly altered, leaving some areas without an adequate blood supply. By reducing the amount of energy needed by the brain areas at risk, they can be protected during the healing.

Leonard underwent heart surgery to fix one artery that was completely blocked and another that was 80 percent obstructed. The heart surgeon was pleased with the result. "His body is in good shape right now. It's just a matter of waiting to see how soon he'll come out of the coma," he told Winifred.

A day and a half later Leonard emerged from his coma and entered the gray zone. "The news is not good," said the doctor. "Leonard's brain is severely damaged. He's in a vegetative state, and he's probably not going to make it."

The events of May 2010 set Leonard and Winifred on a collision course with my team at Western's Brain and Mind Institute. It was only a matter of time. . . .

≈

It was Damian Cruse, our resident EEG (electroencephalography) genius, who had the brilliant idea of buying a Jeep to go visit patients and, even more brilliantly, dubbed this mobile lab the EEJeep. It was the next step in our quest to plumb the depths of consciousness, and it was exactly what I'd been looking for: a mobile solution that would allow us to reach out to gray-zone patients everywhere and put them back into contact with their families. It was a way to bring humans and machines together, melding the organic with the artificial, bonding synapses with silicon. In a move that felt like a hangover from my whacky days at the Unit in Cambridge, I commissioned Wes Kinghorn, an artistic friend, to design a logo for the hood, the rear hatch, and the two front doors. "Make it look like *Jurassic Park*, but not so close that we get sued," I said.

The result was fantastic! The unforgettable *T. rex* skeleton was replaced with a cartoon brain. The trademark red-on-yellow design was switched to purple and white—Western's colors. And the jungle profile was cleverly switched to a profile of the university with its two majestic towers. Riding around town that summer, heads turned. "Is that . . . ? What is *that*?"

The Jeep was simply an elaborate delivery mechanism for our new "secret weapon"—portable EEG brain-imaging gear. The technology is different from MRI or PET, but the goal is the same: to detect consciousness and, when possible, communicate with unresponsive patients. By finding a way to make our equipment mobile, we could finally visit patients such as Leonard in their homes, care facilities, and hospitals. The implications were huge, not only for brain injury, but for neurodegenerative conditions such as Parkinson's disease and Alzheimer's disease, debilitating conditions that lead to incapacity of mind and body, conditions that are becoming more and more common as life expectancy increases.

The arguments were simple—fMRI, the incredible technology that had given us our first opportunity to open a window on consciousness and peer inside, is expensive and by no means portable. The costs of transporting patients to the scanner include ambulance fees, hotel bills for relatives, nurses' salaries, and several days in an expensive care facility—and that's before we've paid for the scans themselves. Developing technologies that would allow day-to-day communication, not in a scanner but in the home, would open up a whole new world of opportunities. More patients could be scanned and costs would be drastically reduced, and as a result, our efforts to explore the gray zone and confront, in the most fundamental terms, what makes us what we are would be accelerated beyond our wildest dreams.

≈

In the summer of 2015, Damian, Laura, and I piled into our newly acquired Jeep and took a short one-hour drive from London to Brantford, a pleasant city of one hundred thousand residents in southwestern Ontario. We were headed to see Winifred and Leonard.

Leonard's predicament was preying on my mind. I'd last seen them both in my office some months earlier, which was unusual. I typically see patients and their family members at our scanner center, at home, or in hospitals or care facilities. On this occasion, Winifred and Leonard were visiting their daughter, who was a student at Western, and they had asked to visit. It always strikes me as incredible that people in the gray zone—nonresponsive as they are, vegetative perhaps, and certainly highly dependent on their caregivers—can travel huge distances, go to the movies (with assistance), watch TV, and sit at the family table at Thanksgiving. All the while it's never quite clear whether they are there.

In my office, the atmosphere had been upbeat and frenetic—almost jovial. Winifred enthusiastically imparted the latest news about Leonard. His bedsores had healed and he was becoming more responsive by the day. He was even pleased to see me, she said. But our news was different—we had looked at Leonard's fMRI scans—and laying this out to Winifred and Leonard was not easy.

Laura and I batted the facts back and forth—we've done it many times, and she's a perfect shield to my reluctant axe. In our recent examination we'd found no evidence from his behavior that Leonard was aware of where he was, who he was, or anything else going on around him. Even imagining a game of tennis, our "gold standard," had failed us on this occasion. Despite his lying still in the scanner for more than two hours, Leonard's brain showed no signs of real life. No messages from the gray zone.

Winifred listened to what I had to say but was keen to add color to our observations. We'd noticed that Leonard was looking physically healthier than when we'd last seen him. Winifred added that he was more responsive and enjoying a day away from the usual routine. We were glad to see that his leg infection had cleared up. Winifred was happy about that too; it enabled Leonard to be much more mobile than before. I'm not saying that Winifred was being disingenuous. She was sincere beyond belief, and for sure she had spent a whole lot more time with Leonard than we had. She certainly knew how to spot signs of improvement where we might not. Was Winifred imputing aspects of consciousness that Leonard could not possibly have had? I wondered. Was some aspect of his person still there? Perhaps she was connecting with some part of him that was entirely unavailable to the rest of us. To determine whether this was so, we'd have to get into Leonard's home and into Leonard's head.

≈

That was how Damian, Laura, and I found ourselves barreling along Highway 401 toward Brantford in the summer of 2015. We pulled up at a sprawling bungalow facing out across a quiet road to open cornfields bathed in sunshine. It was a gorgeous day. Winifred came bounding out of the house to greet us. She'd just arrived home with Leonard and was pushing him up the system of metal ramps that had been constructed to get him through the garage, over the steps, and into a side door. "Welcome, welcome, *welcome*!" she exclaimed.

Damian unpacked the EEJeep, shuttling into the house the sleek black flight cases that we use to transport and protect our EEG gear. Winifred attended to Leonard. I stood looking out over the glistening corn, replaying in my mind that day in my office with the two of them. Would today be different? Would we find better news? Would I have to make yet another stark assessment? The stakes had changed since we had last seen Leonard. We had better tests, better ways of analyzing our data, and more sensitive tools for detecting awareness. I very much wanted better news.

Sitting in the corner of the living room in his wheelchair, Leonard loomed large. "I have been feverishly working," Winifred said. "He has made small but significant steps. He's starting to smile!"

Winifred told us that the night before Leonard's cardiac arrest they had been planning a vacation in India to see Leonard's family, who had retired to Goa. "We were going to book a flight. But we were watching *Dancing with the Stars*, and by the time it was over, it was late, and we decided to wait until the next day to do it. But tomorrow never came."

Winifred asked Leonard to drink water from a plastic cup through a straw. "You have to *sip* it," she scolded. She gently rubbed his cheek and throat. "If you show me you can swallow, I'll give you more. You have to *show* me. I'm trying to wake you up. One more sip and I'm done. I want to see you *swallow*." Her energy was astounding. "No going to sleep. You must stay awake!" She entwined her fingers in his. "Did you see that sigh?" The question was aimed at me.

I was at a loss as to how to respond. I did see the sigh, but was it a conscious response to Winifred's cajoling or was it just an automatic, subconscious reaction that meant nothing? Watching her interact with Leonard, I began to question what makes a person a person. Clearly, Leonard was there, sitting in front of me, but some critical part of Leonard's *being* was not. Not to me, anyway. But to Winifred, Leonard was there, all of Leonard, even the parts that were completely invisible to the rest of us. He lived on in his wife. It almost seemed as if she were carrying his consciousness, keeping it alive and present until such time that he could, once again, carry it himself.

Damian asked for some water to fill a small bowl that we had brought along as part of the gear. He pulled out one of our EEG caps and dumped it straight into the bowl, much the way you would dump a big handful of spaghetti into boiling water. Water conducts electricity well, and by getting all the electrodes soaking wet, Damian was ensuring that he would get a good electrical signal from Leonard's scalp.

Our EEG cap has 128 electrodes attached to bits of rubber mesh and looks like a big hairnet. Each electrode has a wire running from it, which all get grouped and plugged into a device much like a hi-fi amplifier, a metal unit about one foot square. The amplifier is hooked into a top-of-the-line laptop computer, which we go out and buy much the way you would. It's usually an Apple or a Dell.

EEG works in a rather different way from fMRI. When neurons become active, or "fire," they emit electrical activity—a tiny fluctuation in voltage that is detectable at the scalp. It's generally not possible to measure the electrical activity of a single neuron—

not unless you implant electrodes directly into the brain (which requires costly and risky neurosurgery). Neurons fire in bundles, however, and the overall change in voltage produced by a group of them can be detected, even outside the skull. The tiny signal has to be put through an amplifier to make sense, but it is detectable nonetheless.

When we talk about a part of the brain becoming "active" (for example, when the premotor cortex lights up as you imagine playing a game of tennis), we mean that many neurons in that general region are firing more than they were before you started imagining playing tennis. This produces a change in electrical activity that we can detect at the surface of the head with our EEG electrodes. The system isn't perfect because of the "inverse problem," which means that the electrical signal arriving at an electrode on the head could have come from any combination of different neurons firing. The neurons directly underneath that electrode might well be firing, but other neurons, farther away, may also be contributing to that signal. The number and combination of possible contributing neurons is virtually infinite, meaning that it is impossible to relate an EEG signal to a precise location in the brain. Some improvements are being made. Combining EEG with fMRI can be helpful, for example, along with some new statistical techniques being developed, but EEG is still stuck with the inverse problem.

EEG is also limited because all the electrodes are attached to the scalp, meaning that most of the activity that can be detected is close to the brain's surface. There's no chance of detecting activity in the parahippocampal place area, for example, which is used for location memory. It's underneath the brain—too far from the outer surface.

Damian pulled the dripping EEG net from the bowl and said, "We usually have about half an hour or forty-five minutes of good signal until the sponges dry out."

He carefully fitted the EEG net onto Leonard's head. Water ran down Leonard's face as Damian wiggled the net back and forth until it was snug.

"I know home is good for him," said Winifred. "He's opening up his fingers. Does that mean he's feeling and responding? To me it means *something* is connecting. He gets massage. But if he's not in the mood, he'll wince and frown. If you engage him during the day, he's exhausted and sleeps through the night."

Again I was struck by how Winifred attributed thoughts, feelings, and attitudes to Leonard, emotions that she could undoubtedly *feel* whether or not Leonard felt them himself. The gray zone teaches us that consciousness is not an all-or-nothing affair. It's not just on or off, black or white. There are many shades of gray.

"Okay, buddy, I'm going to put earphones in your ears," Damian said.

"We need to use that mind of yours!" Winifred exclaimed.

Damian plugged in the amplifier, flipped open his laptop, and fired up the program, saying, "We all need to be quiet now to make sure that Leonard is not going to be distracted." The room fell silent as we watched Leonard intently.

The kind of brain imaging we're doing with the EEJeep has been made possible by dramatic increases in computing speed and portability. We can analyze massive amounts of data in real time, asking questions and interpreting responses while the patient is under the net. The EEG system is much more streamlined than it used to be. When we scanned Kate back in 1997, we had to write most of the code for data analysis ourselves. It wasn't easy. The MATLAB software didn't have a fancy interface like, say, that of MS Word. Anyone who didn't have scientific training in computing wouldn't have had a clue how to use it. There were no manuals, no help systems; we improvised. Dial forward to today, and much has changed. Software for analyzing EEG data isn't exactly available off the shelf at Best Buy, but it is widely available within the scientific community, and people often share code.

Leonard sat quietly, listening to the sounds that we piped through his headphones. We couldn't hear what Leonard was hearing, and we had no idea whether Leonard could either. We just had to wait to see what the data told us. What was being played through the headphones was a cornucopia of words and phrases dreamed up by Damian, carefully designed to unearth what might be going on in Leonard's brain. The words were played in pairs. Some of the pairs were clearly related to each other, such as *table* and *chair*; others, like *dog* and *chair*, were disparate. This is because of what's known in EEG circles as the N400. When words are presented in pairs, the second word of the pair produces a bigger electrical blip in your brain if it is unrelated to the first. Why this happens is not exactly clear, although we think it's because of a psychological phenomenon known as priming. Priming is related to expectancy: when you hear the word *table*, your brain expects that the next word might be *chair* because *table* and *chair* are often heard together. Similarly, when you hear the word *dog*, the expectation is that the next word is likely to be *cat*. In a sense, the brain is more surprised when *dog* is followed by *chair* than when *table* is followed by *chair*, and this surprise registers as a detectable change in brain activity. That the same word can cause a difference in brain activity, depending on the word that came before it, must mean that our brains have processed the relationship between the two words—our brains must *understand* that *table* and *chair* are more closely related than *dog* and *chair*. Our brains are processing *meaning*. Something similar happens when we hear a sentence such as "The man drove to work in his potato." It generates a bigger change in electrical activity than "The man drove to work in his car." The power of an unexpected ending!

An occasional car whooshed by on the road outside. The room was quiet and still. Almost trancelike. Leonard seemed to drift in and out of sleep. Under Leonard's headphones, Damian's strange poem played:

eagle-falcon; cheetah-trailer
crow-starling; iguana-sweater
basement-cellar; mandarin-fence
dagger-knife; leotards-camel

In addition to several hundred pairs of related and unrelated words, we played Leonard the same sort of signal-correlated noise that we had used to test Debbie more than fifteen years earlier—short bursts of carefully controlled noise, like static from an old radio that leaps out at you from between the stations as you turn the dial. By assessing whether the electrical activity was different for related and unrelated pairs of words, and whether words produced a change that was different from signal-correlated noise, we hoped to discover exactly what Leonard's brain was still capable of. It wasn't a million miles away from what we had done to Debbie, except now we were doing it using equipment that was about sixty times less expensive, and we were doing it *in Leonard's own living room.*

≈

The EEG testing seemed to take a long time, but finally we were done. Damian pushed his fingers up between the net and both sides of Leonard's head, pulling the net apart and lifting it straight up. Leonard didn't budge. He'd hardly moved through the entire process. This was important: the less movement, the more likely we were to get good, clean data from Leonard's brain.

Damian packed up, and Winifred and I walked out to the EEJeep together. I noticed a gray Ford Mustang convertible parked in the driveway. It somehow seemed out of character for Winifred, and I asked her about it.

"It was Leonard's pride and joy. I still take him for a ride. I can see he enjoys it!"

Before we left, Winifred mentioned that her goal was to follow through on the plans they were making the night before Leonard stopped breathing. "I *still* want to take him to Goa. I'm hopeful we can go. My goal is to get him back there. When I talk about it, his face lights up. His eyes get *wide*. He has not forgotten our plans."

Winifred asked about my book and told me to let her know if she could do anything to help. "It's been my passion since day one," she said. "People like Leonard need a voice. If your tests aren't getting something out of Leonard's brain, you need to improve your tests!"

Winifred's words echoed inside me as we hit the 401 back to London, Ontario. "People like Leonard need a voice," she had said. She was that voice. She reminded me that gray-zone science was about affirming the value of every life. The quest to uncover the ubiquitous nature of consciousness inevitably returns to the many ways that each of us is unique. We each contain whole worlds inside our heads, worlds that are built on a lifetime of experience. And for the most part, those worlds are ours alone.

≈

A month or so later I gave Winifred a call from my office at Western. As ever, Laura sat beside me. We'd spent twenty minutes poring over the results of Leonard's EEG.

"How's Leonard doing?" I asked.

Winifred was as chirpy as ever. "He's improving every day! He's even making sounds to tell me that he's feeling better than last week!"

It was impossible not to get caught up in her boundless optimism. "That's fantastic. Well, unfortunately, we don't have anything new to tell you."

Try as we might, we could find no evidence from Leonard's EEG that his brain could tell the difference between words and nonword sounds. "I was really glad to see Leonard's physical condition improving," I said, trying to sound upbeat.

"You see!" Winifred excitedly exclaimed. "I told you that he was getting better and better every day!"

I promised to stay in touch and keep Leonard at the top of my list when it came time to road test our next big idea. As I put the phone down, I couldn't help but wonder whether Winifred was, in fact, right all along about Leonard. All those small changes, all those physical improvements, all those subtle signs.

Perhaps Leonard was gradually making his way back. But back to where? At what point on the trajectory from nothingness to full consciousness do you start to become yourself again? In my explorations into the gray zone, I'd encountered so many people like Leonard: there appeared to be *something* there, at least in the hearts and minds of those who loved them. Some part of these people persisted, beyond the physicality of their bodies and brains. A part of them that was immeasurable, beyond our reach, off the radar. But what was it?

I knew Winifred was right. We did need better tests. We had to keep refining our methods, tweaking our algorithms, discovering new ways to keep making contact. The primacy of human connection—that was what she was insisting upon. That was what was most important. It transcended all the ingenious tests and data points and mind-blowing technology.

I found myself hoping that one day Winifred would fulfill the promise she and Leonard had made to each other the fateful night when Leonard tumbled into the gray zone. With his wife beside him, he would return to India, where their journey had begun so many years ago. The singular arc of their lives would come full circle. Winifred would take her husband home.

CHAPTER FIFTEEN

READING MINDS

≈

> The saddest aspect of life right now is that science gathers knowledge faster than society gathers wisdom.
>
> —Isaac Asimov

Sitting in the smallest, and quite possibly the most quintessentially French, five-star restaurant in Paris recently, I couldn't help but marvel at how much gray-zone science had rippled outward to embrace our quest to understand consciousness itself. L'Hotel is nestled in the heart of the Left Bank of the Seine and has been serving culinary miracles for two centuries. It was early July and a beautifully warm Paris evening. The street outside was filled with the hustle and bustle of Parisians making their way home from work or heading out for the evening. Inside the restaurant, red and black velvet chairs were scattered around small round tables, each adorned with an array of large wineglasses atop crisp white cloths.

My friend and colleague Tim Bayne ordered snails. Tim is a professor of philosophy from New Zealand whose research focuses, among other matters, on the nature of cognition, how it relates to language, whether we have control of our thoughts, and whether modes of thought are culturally specific. He's written extensively about gray-zone science and has always been an enthusiastic supporter of our research.

Opposite Tim and I sat Axel Cleeremans, a Belgian psychologist and world-renowned expert in how learning—with and without consciousness—happens in the brain. Axel and Tim, together with their colleague Patrick Wilken, have produced the excellent *Oxford Companion to Consciousness.* Rounding out our little group was Sid Kouider, a cognitive neuroscientist from Paris who does EEG research on young infants to try to understand how and when consciousness emerges. He, like the others in our group, is obsessed with liminal states, the elusive boundary between brain and mind, being and nonbeing, consciousness and the abyss.

Our first course arrived: snails from the Seine stewed in royal-pink garlic. The dish was exquisitely presented and clearly designed to offer us a glimpse into the chef's approach to the art of cuisine. Soon the laughter flowed as easily as the wine. We were celebrating! Along with a number of our colleagues, we had recently been successful in securing money from the Canadian Institute for Advanced Research (CIFAR) to run a program of meetings on the subject of brain, mind, and consciousness—two or three intense workshops a year in international locations of our choice.

The previous year, CIFAR had launched a global call, asking for "Four Ideas to Change the World," and had received 262 applications from twenty-eight countries on five continents. Our program on brain, mind, and consciousness was one of only four to be funded internationally.

That night in Paris, the four of us focused on the promise of new technologies to help us finally begin to get insights into what parts of the brain need to be working, or connected, for consciousness to emerge. My team's work with the Alfred Hitchcock movie in patients closely paralleled Sid's recent research in five-, twelve-, and fifteen-month-old infants. He and his colleagues had recently shown that an EEG signal that is indicative of consciousness in adults is already present in these tiny infants, in much the same way that we'd shown that an fMRI response related to consciousness is present in some of our patients during *Bang! You're Dead*.

Axel and Tim were convinced, but the four of us debated the findings nonetheless. So-called physiological "signatures" of consciousness—whether they're derived from EEG, fMRI, or any other method—invariably spark serious debate because people rarely agree on what *exactly* they mean. Do those squiggly lines on the EEG trace represent consciousness itself, or are they merely the neural signposts telling us that consciousness is present? Does it matter? If the signposts are there, then we know that the patient (or the infant) is conscious, whether or not we have accessed their consciousness itself.

By analogy, imagine that we were trying to hunt down the physiological "signature" of a particular memory—say, where and how your memory for the title of this book is stored. In the neuropsychological literature, this elusive brain signature is often referred to as the engram—I say "elusive" because we still don't know where or how memories are stored in the brain. We could monitor your brain, using EEG or fMRI, as you sought to recall the title of this book, and no doubt we would see a series of squiggly lines or colored blobs at the exact moment that the words *Into the Gray Zone* popped into your head. But what does that signature represent? Is it the engram? Probably not. Rather than representing the essence of the memory itself, what we would likely be looking at would be the brain processes for delving into your memory to retrieve a previously stored item, the experience of finding out that you know something that you weren't previously sure you knew, or a multitude of other possibilities associated with the *experience of retrieving a memory*. And consciousness is no different. When we seek to measure consciousness, we invariably find that we are measuring brain changes associated with the *experience of being conscious* rather than consciousness itself.

This lively and enjoyable discussion in the best of surroundings was made more so by the constant flow of exquisitely prepared food and fine wine. As the evening progressed, and the wine took hold, we contemplated a future where technology might so progress that the division between the biological and the technological would blur. Our work was pushing us up against an imminent future when telepathy would be possible, not through a magical melding of two minds, but through technology: supercomputers in the palms of our hands that can decode our thoughts and convey them to another being.

Twenty years from now, so-called brain-computer interfaces, or BCIs, will be as commonplace as smartphones, flatscreen TVs, and iPads. A BCI takes a reading of a brain response, analyzes it, and turns it into an action that reflects the user's intention. That action might be as simple as moving a cursor across a computer screen or as complex as manipulating a robot arm to bring a cup of coffee to your lips. Interfaces based on EEG technology already exist. One system presents people with a screen display of letters from *A* to *Z* and asks them to focus their attention on specific letters. Columns and rows of letters flash in a seemingly random order. As the letter that the person wants to convey flashes, and the person focuses attention on it, a tiny electrical signal known as the P300 is emitted by the brain—the brain's equivalent of an "Aha!" moment. Something we've been expecting has finally occurred. EEG detects that brain signal, and via some fairly sophisticated analyses, software can decode what letter flashed at the exact moment that the signal was emitted and then type that letter on a computer screen. It's not the fastest way of communicating—it takes several seconds to type each letter—but with some training most of us can use it to spell out a phrase such as "Hey! I'm conscious" in minutes.

Many challenges are still to be overcome before these systems will allow patients in the gray zone to communicate routinely with the outside world. To use the speller described above, you need to be able to focus your attention on one letter at a time, and that means you have to be able to fix your gaze, not something most people in the gray zone can do. But we and others are designing new systems based on sounds rather than visual cues. You'll simply have to listen out for the letter you're trying to convey.

As we saw in the last chapter, EEG itself has some technical limitations, caused in part because the tiny electrical signals emitted by the brain have to travel through the skull and the scalp before reaching the detecting electrodes. One way around this is to place the electrodes directly on the brain's surface—a complex neurosurgical procedure for sure, but one that can produce miraculous results. At the Brown Institute for Brain Science in Providence, Rhode Island, Cathy Hutchinson, forty-three, had been unable to move her arms or legs for fifteen years, yet she was taught to control a robot arm using just her brain. A sensor implanted in her brain and connected to a decoder turned her thoughts into instructions to move the robotic arm. Cathy, a post office employee, was a single mother of two when she suffered a catastrophic brain-stem stroke in 1996. The stroke left her locked in—unable to move any of her limbs and unable to speak. But with the aid of the advanced BCI, Cathy was able to steer a robotic arm toward a bottle, pick it up, and drink her morning coffee.

This new technology may soon allow people in the gray zone to take online courses, type e-mails, hold conversations, and express their innermost feelings. Challenges remain, both technical and ethical. Brain surgery is risky, and implanting electrodes on the surface of brains should not be undertaken casually. Cathy Hutchinson could control her eyes, and with some clever engineering she could slowly pick out letters on a keyboard, allowing her to communicate that she was conscious and consented to the surgery. Presumably, after fifteen years with no mobility in her arms or legs, she thought it was a risk worth taking.

Can you imagine the implications for people in the gray zone, or in the advanced stages of Alzheimer's or Parkinson's disease? We're poised on the brink of a world where electrodes implanted into the brain may allow patients who have not been able to communicate their wishes for decades to reassert their autonomy, to take control of their lives and once again guide their own destinies. Those who have been voiceless will speak again, those who have been unable to move will move again, and those who we thought were gone forever will be brought back into the here and now, to exercise their right to be treated as real *people* with plans for their futures and memories of their pasts.

≈

The gray-zone technology of reading minds has found a fascinating application in an unlikely field: forensic investigation. In 2015, my team came across a man in his twenties, Dan, who had been shot through the head, in Sarnia, Ontario. He was critically ill and on life support at a local hospital. It was a rare event: Ontario is generally a safe and peaceful place. Dan was left with a catastrophic brain injury, alive but nonresponsive. The bullet entered his forehead right between the eyes, traveled through his brain, and exited between his parietal cortex and his temporal lobe. No one knew who shot him. What if we scanned him, established that he was actually conscious, and then asked him who had done it?

A recent episode of the TNT TV show *Perception* used our research to build a plotline involving almost exactly this scenario (www.intothegrayzone.com/perception). The technology exists. A successful interview of a crime victim in the gray zone *can* be accomplished. It would take a bit longer than it did in the *Perception* episode, and the main characters might be a little less glamorous, but where the victim is the best source of the truth, fMRI can establish who has committed a heinous crime.

Was Dan that victim? We rushed to get permission to scan him. Major ethical hurdles had to be overcome. Why were we doing this? It obviously wasn't purely for research purposes. It wasn't clinical either. It was to solve a crime! How could we persuade our ethics committee to let us go forward? Who would give consent? Who was Dan's substitute decision maker? What if the substitute decision maker was the perpetrator of the crime? How would we know?

Our admittedly vague plan was to get a list of all of Dan's friends and associates, and then, with Dan in the scanner, start by asking him to imagine playing a game of tennis if he knew who did this to him. If he answered yes, we'd start going down the list: "Was it Johnny? Imagine playing tennis for yes, imagine moving around your house for no." Then, "Was it Dave?" And so on. We became tremendously excited. It could work! Our research methods could solve a crime.

Then Dan recovered. A few days into our deliberations, he became conscious. He was able to raise his hand on command. We'd missed our chance to reach out to him and see what he could have told us using just his brain. It was fortunate for Dan, but part of me was disappointed.

Dan wasn't the patient to show us that fMRI can contribute

to forensic science, but sooner or later another patient will come along. We'll get to someone who is unable to communicate by normal means but whose mind will be readable with our rapidly advancing technology. It hasn't happened yet, but it will.

≈

The questions we have tackled and the technologies we have developed with gray-zone science have opened up a whole new world of scientific possibilities. Our Alfred Hitchcock experiments may be able to give us some answers to what's going on in the brains of patients with cognitively debilitating neurodegenerative conditions such as Alzheimer's disease. When people with Alzheimer's watch a classic thriller from the Master of Suspense, is their experience like yours and mine? Or is it more like an infant's—sounds and visual cues echoing through the brain but the subtleties of the plot not registering? If so, can we develop assistive technologies and therapies that are tailored to each patient's actual experience of the world rather than the experience that we assume they must be having as we look on as observers from the outside? The recent documentary *Alive Inside*, which won the Audience Award at the 2014 Sundance Film Festival, chronicles the astonishing experiences of several people with Alzheimer's disease whose lives were turned around when they were played music that they had known and loved. Each patient made a personal connection with *their* music, with *their* past, with some aspect of their being that those close to them thought had been lost. The film beautifully renders how music has the capacity to reawaken our conscious selves and uncover the deepest parts of our humanity.

≈

In addition to research into the deterioration of consciousness in conditions such as Alzheimer's, promising work is being done on what some call animal consciousness. *Are* other animals conscious? Most people tend to think that apes, dogs, and other higher primates possess some form of consciousness, but clearly it's not exactly like ours. We know that the scaffolds of consciousness are there, but they are not as integrated and established as in the human brain. Koko, the western lowland gorilla who was born at the San Francisco zoo, has been able to learn the meaning of thousands of hand signs, as well as many English words. Yet most agree that she does not use grammar or syntax, and her linguistic abilities do not exceed those of a young human child. Likewise, many animals, including dogs, can be taught complex sequences of actions in response to commands, yet their behavior is afterward bound by the sequence—they can't spontaneously elaborate and improvise on the behavior (for example, performing the sequence in reverse) the way a human being can.

Seated across from one another at our L'Hotel dinner table, Tim, Axel, Sid, and I pondered the subject of animal consciousness as it relates to adult consciousness, infant consciousness, and machine consciousness. It always amazes me that most scientists, however well versed they are in the world of consciousness research, still resort to discussing the "consciousness" of their own pets. The truth about what these creatures are capable of is often more complicated than you might think.

Although other species may be capable of rudimentary forms of thought including deception, the mature manifestations of these phenomena appear to be ours alone. Can the members of other species think about their own consciousness as we do, traveling backward in time to ponder their past and forward to plan their futures? We can't say with certainty, but we would all agree, I think, that emotional experience is not uniquely human.

Few dog owners would say that their pets don't express strong emotions. But the complexity of human emotion and our ability to communicate our feelings via art or music is surely unique. And in other species, consciousness does not appear to involve as much interaction with the minds of others. From infancy onward we invest much of our time and energy in attempting to identify what others are thinking, what their motives might be, whether they love us or not, and what they're likely to do next. Whether or not you know it, you spend much of your life trying to understand the conscious states of other people and trying to communicate—or hide—your own.

Emerging technologies will undoubtedly one day allow us to *read* the minds of others. Not in the rudimentary sense that we do already—decoding yes and no responses based on changes in fMRI activity—but in the sense of interpreting and understanding exactly what another person is thinking based solely on some sort of readout from his or her brain. The ethical conundrums that this will produce will be immense in business, politics, and advertising; there will be an insatiable (and sometimes sinister) appetite for access to the thoughts of others. The way that the world operates will radically change, much as it has changed since the advent of the Internet and the World Wide Web. But we will adapt as a species, and these changes will just become the way things are: the tools that our children will use from birth, and the technologies that will define the blueprint for the generations that follow.

The advent of increasingly autonomous machines capable of initiating their own courses of action will inevitably require that they be imbued with a sense of moral responsibility, one that is in many ways superior to our own. We humans have an unusual (and at times unnerving) knack for doing things *just because we want to.* These may be wrong, immoral, illegal, or illogical, but still we often choose to do them. What in our DNA allows us to go beyond what is logically right and do what is patently wrong? Discovering the origin in ourselves of this tendency to waywardness may help us to safeguard against the same impulses in machines.

As we mull the nature of consciousness and the ability to act on our thoughts (a capacity that some call agency and that many gray-zone patients often lack), it is worth considering: Do we have free will? While many great minds have wrestled with this thorny problem, the answer may be even more complex than we think. Winifred and Leonard show us how our consciousness frequently spills over into the lives of others. Rarely can we fully describe or understand ourselves without reference to our relationships and the impact that we, as conscious beings, have had on the world around us. We are our brains, but we're also the memories, attitudes, opinions, and emotions that we imbue in others. Even in death we often continue to inspire, mold, and affect the lives of the people we leave behind.

This phenomenon manifests itself perhaps most clearly in what some call collective consciousness. We live in overlapping groups of families, communities, and nations. Other clusters cut across these divisions, such as religious organizations and sports clubs. Because individuals within each of these cells continually act upon and influence one another, these groups possess a kind of agency—the capacity to make decisions, think, judge, act, organize, and reorganize. They're even able to reflect on their roles as agents, possessing a sort of "will" in the form of shared beliefs, moral attitudes, traditions, and customs.

Collective consciousness emerges out of the interactions between each of our brains, escalating upward as people, families, communities, and even nations touch. Collective consciousness is key to our humanity, our sense of *being something* over and above atomized individuals. Its trickle-down effect sculpts our beliefs and fuels our prejudices. It is the foundation of shared conscious experiences, from the joint rapture of a fulfilling sexual encounter to the spontaneous, synchronized behavior of one hundred thousand like-minded individuals joyously performing a "wave" at the Olympics.

Collective consciousness has features in common with what some have termed universal or cosmic consciousness. Universal consciousness is said to be "an infinite, eternal ocean of intelligent energy. Each of us, each soul, each individual point of consciousness, is a drop in that ocean. Where one drop ends and another begins is impossible to determine because there is no separation in the unified field of energy." As a metaphor, this description has some appeal: the consciousness that we each possess is a "drop" in the collective ocean. Exactly how we each contribute to the whole is certainly impossible to determine, in large part because much of life emerges in a way that transcends each of us and our individual contributions. Life is a dazzling improvisation. We make it up *collectively* as we go along. That's what makes being alive so interesting!

As I raised a glass to the future of gray-zone science, it occurred to me that even the course of a lively conversation between four friends in a Parisian restaurant is impossible to predict. Each conscious mind subtly influences the disposition of the group. Ideas germinate and are modified, embellished, discarded, or embraced. The possibilities multiply into the future, radiating from their genesis, creating and responding to a world that is relentlessly *becoming*.

My view is that we do not need concepts such as "unified fields of energy" or "infinite, eternal oceans" to explain the emergence of consciousness. All we need is the brain itself. Each and every one of our 100 billion neurons has a role to play. Each of our neurons is not just a transistor or a switch. It is a tiny engine of decision making, "deciding" when and when not to fire. Countless decisions are being made inside us moment by moment. As we've seen, a neuron in the fusiform gyrus may respond to one face and not another. A "place cell" in the parahippocampal gyrus may respond to one place and not another. And sometimes, neurons in the brain stem or the thalamus may not respond at all, plunging us into the gray zone.

Each of us around that dinner table in Paris, along with many thousands of our colleagues around the world, believes that these tiny decision makers, and their hundreds of billions of interconnections, are the basis for the emergence of consciousness in the form of our thoughts, feelings, emotions, memories, and plans. Every neuron is part of the scaffold of consciousness, just as each one of us is part of the fabric of society. Some contribute more than others. But working with people in the gray zone has shown me that it's vitally important to remember that just by *being*, each of us contributes to the emergent whole.

I am convinced that consciousness is reducible to the connections between neurons firing at one another. Yet in its most elaborated form it is the part of being human that we treasure most—our sense of self, of agency, of *being* something. It's no wonder that it is so hard to comprehend. My explorations into the gray zone have taught me that consciousness is not inexplicable, mystical, or metaphysical. *Strange*, perhaps. Even magical. Especially in the ways it spills out from us into the lives of others. Bigger than any one of us, it carries us along in a ceaseless flux to destinations we can't even begin to comprehend.

Twenty years ago, many people dismissed our quixotic quest to read the minds of patients lost in the gray zone. Yet soon such decoding will be commonplace, available to millions all over the world. This is the magic of science, pulling the future into the past, chipping away at every problem until we can hardly believe the progress we've made as new realms of insight and understanding unfold before us. Gray-zone science has brought us a long way since 1997, when we first scanned Kate. Ultimately, it promises to reveal the secrets of the universe that each of us, incredibly, carries inside our heads.

EPILOGUE

The first chapter of my explorations into the gray zone came to a strange and unexpected close in May 2015 when Maureen, quite suddenly, died. I had kept in touch with Phil—I had last seen him seven months earlier when he and I met up in Edinburgh for a beer. He told me at that time Maureen was still medically stable, still living in the nursing home, and still being lovingly cared for by her parents and her family. On the day she died, I was flying to New York City to talk to publishers about this book. Phil contacted me on Facebook that same day: "Maureen died at 9:20 this morning after struggling with a chest infection for two days. She went quickly. . . . Thought you would like to know."

Spooky was the word that came to mind when I thought of the timing of her death. As I traipsed up and down Fifth Avenue peddling my story, I had to explain again and again to publishers that Maureen had, coincidentally, just died. I felt like the Ancient Mariner! Maureen had haunted this work from its inception, just as she had haunted my life for almost two decades, yet by exiting the gray zone right then, she was continuing to do what she had always done, impacting my life in strange and unpredictable ways, always voicing her opinion, always having the last word. Only now she was doing it from beyond the grave.

I hadn't seen Maureen in over twenty years, yet I was deeply affected by her passing. I acutely felt how much she had influenced the course of my life for two decades, although I had rarely acknowledged that to myself explicitly. Her influence had been hard to quantify and even harder to explain, undoubtedly obfuscated by my conflicting views about our relationship. The rancor of our bruising arguments had long ago receded, but I realized that in some ways I was still responding to her insistence that *caring* should be what's most important.

Beyond the elegant experiments and dazzling technology, the heart of gray-zone science is about finding people who have been lost to us and reconnecting them with the people they love and who love them. Each contact still feels like a miracle. I can hear Maureen laughing as I write this, her eyes crinkling, her sharp wit flashing. *I told you so,* she would say. *You see, it is about caring*. And she would be right. What began as a scientific journey more than twenty years ago, a quest to unlock the mysteries of the human brain, evolved over time into a different kind of journey altogether: a quest to pull people out of the void, to ferry them back from the gray zone, so they can once again take their place among us in the land of the living.

ACKNOWLEDGMENTS

I have written hundreds of acknowledgments in my life, but always perfunctory and always giving thanks to faceless granting agencies (that's not to say I am ungrateful, I just never really know who I am thanking). It's a whole lot more satisfying to acknowledge the contributions of real people.

I'll begin by thanking the real heroes of this story: the hundreds of patients and families who have given me and my research team so much of themselves over the years. Some of your stories made it into the book and some did not, but all of you, without exception, contributed to the process of scientific discovery, and for that I am truly grateful. I am particularly indebted to Kate, to Paul and his son Jeff, to Winifred and her husband, Leonard, and to Margarita and her son Juan, who all took time away from their complicated lives to tell me their stories in their own words. I couldn't have written this book without you all, and I hope I have done your stories justice.

Thanks also to Maureen's parents and her brother Phil, for encouraging me to include her story in the narrative of the book. I was reluctant at first, but the course of my journey would have been incomplete without it. Maureen's generosity of spirit lives on in all three of you.

Over the years, I have been lucky enough to work with a small army of research assistants, technicians, graduate students, and postdoctoral fellows, who all contributed in great and small ways to the success of our scientific quest. Any attempt to name them all would be imperfect and bound to offend someone. Therefore, I will reserve my heartfelt thanks for the main players; those members of my lab who really made this scientific adventure unfold in the way that it did. Thank you (in no particular order), Tristan Bekinschtein, Martin Monti, Davinia Fernández-Espejo, Damian Cruse, Srivas Chennu, Lorina Naci, Loretta Norton, Raechelle Gibson, Laura Gonzalez-Lara, Andrew Peterson, and Beth Parkin. I hope you all had as much fun as I did.

The same is true for the many hundreds of wonderful colleagues that I have collaborated with over the years on the work described in this book; a comprehensive list would be impossible to compile. Nevertheless, I would like to extend my sincere thanks and appreciation to David Menon, Steven Laureys, and Melanie Boly, whose essential contributions to the scientific narrative are described in some detail in these pages. There are, of course, many others who have played significant roles along the way and I am grateful to Ingrid Johnsrude, Matt Davis, Jenni Rodd, John Pickard, Bryan Young, Martin Coleman, Charles Weijer, Rhodri Cusack, and Andrea Soddu. Roger Highfield worked with me on an early proposal for the book and deserves a special mention; without that first push, my story may have never seen the light of day.

I am indebted to my tireless assistant and "bouncer," Dawn, who says she loves her job just often enough for me to believe her, but not so often that I don't. This book project, along with many other aspects of my life, would have fallen apart without her.

Heartfelt thanks to Kenneth Wapner, my collaborator. He helped me find the heart of the story and give it dimension and depth, from the creation of the proposal through every stage of the development of the manuscript. Ken, it was a pleasure and a privilege to work with you and I hope we do it again sometime soon. I am grateful to Gail Ross, my agent, who persuaded me to write *Into the Gray Zone* against my better judgment, and Rick Horgan, my editor, whose guiding hand really made all the difference.

And finally, for those of you who might be searching for hidden meaning in this book's dedication to my son, Jackson, I am not (nor do I fear I am) about to die. But sometimes, working at the flimsy border between life and death, the sheer fragility of every human life is hard to ignore.

—Adrian Owen, February 12, 2017

NOTES

For most of the cases described in this book I have had the invaluable cooperation of the patients and their families, for which I am enormously grateful. For others, sometimes for obvious reasons, I have not. In those few cases, I have changed names, dates, and other unimportant details to maintain privacy.

PROLOGUE

Page 3 *Perhaps most important, we have discovered that 15 to 20 percent of people in the vegetative state who are assumed to have no more awareness than a head of broccoli are fully conscious, although they never respond to any form of external stimulation.*

For further details, see M. M. Monti, A. Vanhaudenhuyse, M. R. Coleman, M. Boly, J. D. Pickard, J-F. L. Tshibanda, A. M. Owen, and S. Laureys, "Willful Modulation of Brain Activity and Communication in Disorders of Consciousness." *New England Journal of Medicine* 362 (2010): 579–89, and D. Cruse, S. Chennu, C. Chatelle, T. A. Bekinschtein, D. Fernandez-Espejo, D. J. Pickard, S. Laureys, and A. M. Owen, "Bedside Detection of Awareness in the Vegetative State," *Lancet* 378 (9809) (2011): 2088–94.

Page 4 *With the help of an assistant and a writing board, he composed* The Diving Bell and the Butterfly, *a memoir, which took two hundred thousand blinks to complete.*

I enthusiastically refer readers to this fascinating and moving book, which I have read several times over the years. J. D. Bauby, *The Diving Bell and the Butterfly* (New York: Vintage, 1998).

CHAPTER 1

Page 16 *Over the next three years, we spent many hours poring over his drawings of the frontal lobes, scribbling little notes about what each area of the brain probably did and designing new tests that would show us how different parts of the brain contributed to memory.*

Versions of several of the memory tests that we developed around this time are now available online at www.cambridgebrainsciences.com.

Page 17 *Much of the early work was confirmatory, but that just added to the excitement. For instance, we'd known for some years that an area on the undersurface of the brain, close to where the temporal lobe and the occipital lobe intersect, is involved in face recognition.*

For some early evidence from patients who have sustained damage to this area resulting in difficulties with face recognition, see J. C. Meadows, "The Anatomical Basis of Prosopagnosia," *Journal of Neurology, Neurosurgery, and Psychiatry* 37 (1974): 489–501.

Page 19 *One of my early successes showed that one area of the frontal lobes was crucial for organizing our memories.*

For further details, see A. M. Owen, A. C. Evans, and M. Petrides, "Evidence for a Two-Stage Model of Spatial Working Memory Processing within the Lateral Frontal Cortex: A Positron Emission Tomography Study," *Cerebral Cortex* 6 (1) (1996): 31–38;

and A. M. Owen, "The Functional Organization of Working Memory Processes within Human Lateral Frontal Cortex: The Contribution of Functional Neuroimaging," *European Journal of Neuroscience* 9 (7) (1997): 1329–39.

Page 19 *This process is an example of what we call* working memory, *which is a special kind of memory that we only need to retain for a limited period.*

Several of the working memory tests that we used in our PET activation studies at the time can now be taken online at www.cambridgebrainsciences.com.

Page 20 *We started to scan patients with Parkinson's disease to try to understand why it is that they, in particular, have problems with working memory.*

For further details, see A. M. Owen, J. Doyon, A. Dagher, A. Sadikot, and A. C. Evans, "Abnormal Basal-Ganglia Outflow in Parkinson's Disease Identified with Positron Emission Tomography: Implications for Higher Cortical Functions," *Brain* 121 (pt. 5) (1998): 949–65.

CHAPTER 2

Page 34 *When our paper describing Kate's extraordinary case came out in the* Lancet, *one of the world's oldest (1823) and best-known medical journals, there was a flurry of media attention.*

D. K. Menon, A. M. Owen, E. Williams, P. S. Minhas, C. M. C. Allen, S. Boniface, and J. D. Pickard, "Cortical Processing in the Persistent Vegetative State," *Lancet* 352 (9123) (1998): 200.

Page 35 *Why did Kate recover?*

Kate was my first experience of such a patient, and, I believe, the first suggestion that any of us working in this field had that a positive brain response in the scanner could indicate the potential for recovery. It would be another twelve years until we were able to publish evidence in the British neurological journal *Brain* from a large group of patients like Kate, showing that a positive response in the brain scanner is a good sign, can herald some kind of recovery, and is therefore a valuable prognostic tool. For further details, see M. R. Coleman, M. H. Davis, J. M. Rodd, T. Robson, A. Ali, J. D. Pickard, and A. M. Owen, "Towards the Routine Use of Brain Imaging to Aid the Clinical Diagnosis of Disorders of Consciousness," *Brain* 132 (2009): 2541–52.

CHAPTER 3

Page 46 *Most of us can listen to and correctly repeat "digit spans" of a five- or six-number sequence.*

If you would like to know how many numbers you can remember, you can test your digit span online at www.cambrigebrainsciences.com.

Page 47 *Through a series of clever studies at the Unit, my former student Daniel Bor showed that this memory recoding, where information is repackaged and organized to make later retrieval easier, is carried out by regions of the brain that have been linked to general intelligence, otherwise known as g, which is measured by tests of IQ.*

For further details, see D. Bor, J. Duncan, and A. M. Owen, "The Role of Spatial Configuration in Tests of Working Memory Explored with Functional Neuroimaging," *Journal of Scandinavian Psychology* 42 (3) (2001): 217-24; D. Bor, J. Duncan, R. J. Wiseman, and A. M. Owen, "Encoding Strategies Dissociates Prefrontal Activity from Working Memory Demand," *Neuron* 37 (2) (2003): 361-67; D. Bor, N. Cumming, C. E. M. Scott, and A. M. Owen, "Prefrontal Cortical Involvement in Verbal Encoding

Strategies," *European Journal of Neuroscience* 19 (12) (2004): 3365-70; D. Bor and A. M. Owen, "A Common Prefrontal-parietal Network for Mnemonic and Mathematical Recoding Strategies within Working Memory," *Cerebral Cortex* 17 (2007): 778-86.

CHAPTER 4

Page 53 *O-15 has a half-life of 122.24 seconds.*

The term *half-life*, when used about a radioisotope such as O-15, refers to the time that it takes for half of the radioactive nuclei in any sample to decay. Thus, after two half-lives (in this case 244.48 seconds), any sample of O-15 will be one fourth its original size, and so on.

Page 58 *When we wrote about Debbie's case in the scientific journal* Neurocase *late that year, we sat firmly on the fence.*

For further details, see A. M. Owen, D. K. Menon, I. S. Johnsrude, D. Bor, S. K. Scott, T. Manly, E. J. Williams, C. Mummery, and J. D. Pickard, "Detecting Residual Cognitive Function in Persistent Vegetative State (PVS)," *Neurocase* 8 (5) (2002): 394–403.

Page 58 *In part, we were responding to the results of a scientific paper that had been published in the* Journal of Cognitive Neuroscience *a year or so after our paper about Kate appeared in the* Lancet.

For further details, see N. D. Schiff, U. Ribary, F. Plum, and R. Llinas, "Words without Mind," *Journal of Cognitive Neuroscience* 11 (1999): 650–56.

Page 58 *Plum was a giant in the field of brain injury.*

Fred Plum coined the terms *persistent vegetative state* and *locked-in syndrome*. His 1966 book, *The Diagnosis of Stupor and Coma*, remains the bible for all of us to this day. For further details, see F. Plum and J. B. Posner, *Diagnosis of Stupor and Coma* (Philadelphia: F. A. Davis, 1966). For the current edition, see J. B. Posner, C. B. Saper, N. D. Schiff, and F. Plum, *Plum and Posner's Diagnosis of Stupor and Coma*, 4th ed. (Oxford, England: Oxford University Press, 2007).

Page 59 *Their brains appeared to be less tightly "connected" than healthy controls, with disorganized or fragmented patterns of overall activity.*

For further details, see S. Laureys, S. Goldman, C. Phillips, P. Van Bogaert, J. Aerts, A. Luxen, G. Franck, and P. Maquet, "Impaired Effective Cortical Connectivity in Vegetative State: Preliminary Investigation Using PET," *Neuroimage* 9 (1999): 377–82.

Page 60 *And the same year that we published our paper about Debbie, Dr. Joe Giacino and colleagues published a landmark paper describing, for the first time, the minimally conscious state.*

For further details, see J. T. Giacino, S. Ashwal, N. Childs, R. Cranford, B. Jennett, D. I. Katz, J. P. Kelly, J. H. Rosenberg, J. Whyte, R. D. Zafonte, and N. D. Zasler, "The Minimally Conscious State: Definition and Diagnostic Criteria," *Neurology* 58 (3) (2002): 349–53.

CHAPTER 5

Page 67 *As Daniel Bor argued in his excellent 2012 book,* The Ravenous Brain.

D. Bor, *The Ravenous Brain: How the New Science of Consciousness Explains Our Insatiable Search for Meaning* (New York: Basic Books, 2012).

Page 72 *As the great Francis Crick, a physicist and molecular biologist, wrote in his seminal 1994 book,* The Astonishing Hypothesis . . .

F. Crick, *The Astonishing Hypothesis: The Scientific Search for the Soul* (New York: Scribner, 1994).

CHAPTER 6

Page 74 *My psycholinguist friends had found a way in—a way to distinguish between a brain that was* understanding *speech and one that was merely* experiencing *it.*

For further details, see M. H. Davis and I. S. Johnsrude, "Hierarchical Processing in Spoken Language Comprehension," *Journal of Neuroscience* 23 (8) (2003): 3423–31.

Page 77 *We had replicated our findings. There could be little doubt that Kevin's brain was processing* meaning.

For further details, see A. M. Owen, M. R. Coleman, D. K. Menon, I. S. Johnsrude, J. M. Rodd, M. H. Davis, K. Taylor, and J. D. Pickard, "Residual Auditory Function in Persistent Vegetative State: A Combined PET and fMRI Study," *Neuropsychological Rehabilitation* 15 (3–4) (2005): 290–306.

Page 82 *It's possible with fMRI to see how a single sentence such as "The boy was frightened by the loud bark" is decoded into its correct meaning by our brains in milliseconds.*

For further details, see J. M. Rodd, M. H. Davis, and I. S. Johnsrude, "The Neural Mechanisms of Speech Comprehension: fMRI Studies of Semantic Ambiguity," *Cerebral Cortex* 15 (2005): 1261–69.

Page 85 *No experiment like this had ever before been conducted.*

For further details, see A. M. Owen, M. R. Coleman, D. K. Menon, I. S. Johnsrude, J. M. Rodd, M. H. Davis, K. Taylor, and J. D. Pickard, "Residual Auditory Function in Persistent Vegetative State: A Combined PET and fMRI Study," *Neuropsychological Rehabilitation* 15 (3–4) (2005): 290–306.

CHAPTER 7

Page 91 *To paraphrase the famous twentieth-century Canadian neuropsychologist Donald Hebb, "Neurons that fire together, wire together."*

This phrase, or at least a version of it ("neurons wire together if they fire together"), was first used in Siegrid Löwel and Wolf Singer's "Selection of Intrinsic Horizontal Connections in the Visual Cortex by Correlated Neuronal Activity," *Science* 255 (1992): 209–12. However, they were paraphrasing Donald Hebb, a Canadian neuropsychologist known for his work in the field of associative learning. In 1949, Hebb wrote, "When an axon of cell A is near enough to excite a cell B and repeatedly or persistently takes part in firing it, some growth process or metabolic change takes place in one or both cells such that A's efficiency, as one of the cells firing B, is increased." This concept has become known as Hebbian theory, Hebb's rule, Hebb's postulate, or cell assembly theory. For further details, see D. O. Hebb, *The Organization of Behavior* (New York: Wiley & Sons, 1949).

Page 92 *Back in my Maudsley days with Maureen, I showed that patients who had sustained massive damage to the frontal part of their brain were still able to recognize a picture they had seen before, even if they'd only glimpsed it briefly.*

For further details, see A. M. Owen, B. J. Sahakian, J. Semple, C. E. Polkey, and T. W. Robbins, "Visuo-spatial Short Term Recognition Memory and Learning after Temporal Lobe Excisions, Frontal Lobe Excisions or Amygdalo-hippocampectomy in Man," *Neuropsychologia* 33 (1) (1995): 1–24.

Page 94 *If you are inundated with names of aunts and distant cousins, then you may have to bring in the heavy artillery—one particular area within the middle and top half of the frontal lobes known as the dorsolateral frontal cortex.*

Over the years, my team and I have developed specific cognitive tests for assessing how well your dorsolateral frontal cortex might be functioning. Some of these tests are available online at www.cambridgebrainsciences.com. To test your dorsolateral frontal cortex, try the test called "Token Search."

Page 94 *It should come as no surprise then that this region of the brain has also been associated with aspects of general intelligence (g) and performance on IQ tests.*

For further details, see J. Duncan, R. J. Seitz, J. Kolodny, D. Bor, H. Herzog, A. Ahmed, F. N. Newell, and H. Emslie, "A Neural Basis for General Intelligence," *Science* 289 (2000): 457–60; and A. Hampshire, R. Highfield, B. Parkin, and A. M. Owen, "Fractioning Human Intelligence," *Neuron* 76 (6) (2012): 1225–37.

Page 96 *This was interesting in and of itself and made a modest impact on the scientific literature on frontal-lobe function when Anja and I published it in the journal* NeuroImage *two years later.*

For further details, see A. Dove, M. Brett, R. Cusack, and A. M. Owen, "Dissociable Contributions of the Mid-ventrolateral Frontal Cortex and the Medial Temporal-Lobe System to Human Memory," *NeuroImage* 31 (4) (2006): 1790–1801.

CHAPTER 8

Page 101 *Your hippocampus, a sea-horse-shaped structure deep in your brain, has specialized neurons known as place cells, which were first discovered in rats in 1971 by neuroscientist John O'Keefe and his colleagues.*

For further details, see J. O'Keefe and J. Dostrovsky, "The Hippocampus as a Spatial Map. Preliminary Evidence from Unit Activity in the Freely-Moving Rat," *Brain Research* 34 (1) (1971): 171–75.

Page 102 *Close to the hippocampus, in an area of cortex called the parahippocampal gyrus, is a piece of brain tissue that becomes highly active when humans view pictures of places, such as landscapes, city views, or rooms.*

The functions of this region of the brain weren't described in any detail until 1998 by my colleagues Russell Epstein and Nancy Kanwisher, who'd conducted experiments using fMRI in humans. An fMRI study two years earlier, by Geoffrey Aguirre and his colleagues at the University of Pennsylvania, had already pointed the finger at this part of the brain and its potential role in our "mental map" of our environment. Those investigators asked their volunteers to mentally navigate from A to B through a maze that they had previously learned using a virtual-reality system inside the scanner. The researchers found that simply imagining moving through this now-familiar environment activated the parahippocampal gyrus, as scenes and views came to mind. For further details, see G. K. Aguirre, J. A. Detre, D. C. Alsop, and M. D'Esposito, "The Parahippocampus Subserves Topographical Learning in Man," *Cerebral Cortex* 6 (6) (1996): 823–29; and R. Epstein, A. Harris, D. Stanley, and N. Kanwisher, "The Parahippocampal Place Area: Recognition, Navigation, or Encoding?," *Neuron* 23 (1999): 115–25.

Page 104 *Every participant activated an area on the top of the brain known as the premotor cortex.* Every single subject. *All exactly the same.*

For further details, see M. Boly, M. R. Coleman, M. H. Davis, A. Hampshire, D. Bor, G. Moonen, P. A. Maquet, J. D. Pickard, S. Laureys, and A. M. Owen, "When Thoughts Become Actions: An fMRI Paradigm to Study Volitional Brain Activity in Non-Communicative Brain Injured Patients," *Neuroimage* 36 (3) (2007): 979-92.

Page 112 *A single-page article describing our results appeared in* Science *in September 2006.*

For further details, see A. M. Owen, M. R. Coleman, M. H. Davis, M. Boly, S. Laureys, and J. D. Pickard, "Detecting Awareness in the Vegetative State," *Science* 313 (2006): 1402.

Page 119 *In 2000, a case report in the* South African Medical Journal *described a young man who "awakened" within thirty minutes of receiving zolpidem after three years in a vegetative state.*

For further details, see R. P. Clauss, W. M. Güldenpfennig, H. W. Nel, M. M. Sathekge, and R. R. Venkannagari, "Extraordinary Arousal from Semi-Comatose State on Zolpidem," *South African Medical Journal* 90 (1) (2000): 68–72.

Page 120 *A comprehensive recent study by my friend and colleague Steven Laureys in Liège, Belgium, failed to show an improvement in even one of sixty patients with disorders of consciousness who were tested on the drug.*

For further details, see M. Thonnard, O. Gosseries, A. Demertzi, Z. Lugo, A. Vanhaudenhuyse, M. Bruno, C. Chatelle, A. Thibaut, V. Charland-Verville, D. Habbal, C. Schnakers, and S. Laureys, "Effect of Zolpidem in Chronic Disorders of Consciousness: A Prospective Open-Label Study," *Functional Neurology* 28 (4) (2013): 259–64.

CHAPTER 9

Page 122 *Our findings suggested that as many as seven thousand might actually be aware of everything going on around them.*

It is estimated that in the United States aound 5.3 million people are living with a disability related to a traumatic brain injury; in Europe, that figure is close to 7.7 million. According to the

World Health Organization, many of the 15 million people who suffer stroke worldwide each year experience long-term cognitive and physical disabilities. Improvements in roadside medicine and intensive care have led to more people surviving serious brain damage and ending up alive but with no evidence of preserved awareness. Such patients can be found in virtually every city and town with a skilled-nursing facility.

Page 122 *Six years on we would know the answer to this question, but at the time it was puzzling.*

For further details, see A. M. Owen and L. Naci, "Decoding Thoughts in Behaviourally Non-Responsive Patients," in W. Sinnott-Armstrong (ed.), *Finding Consciousness* (Oxford University Press, 2016).

Page 126 *As Rabbi Jack Abramowitz wrote in an interesting blog on the subject in 2014 . . .*

For further details, see http://jewinthecity.com/2014/09/can-you-ever-pull-the-plug-life-support-jewish-law/.

Page 142 *In fact, my colleague Brian Levine at the Rotman Research Institute in Toronto has described a whole new condition known as severely deficient autobiographical memory syndrome.*

For further details, see D. J. Palombo, C. Alain, H. Södurland, W. Khuu, and B. Levine, "Severely Deficient Autobiographical Memory (SDAM) in Healthy Adults: A New Mnemonic Syndrome," *Neuropsychologia* 72 (2015): 105–18.

Page 145 *When our paper describing John's case was published, once again my lab was deluged with frenzied media attention.*

For further details, see M. M. Monti, A. Vanhaudenhuyse, M. R. Coleman, M. Boly, J. D. Pickard, J-F. L. Tshibanda, A. M. Owen, and S. Laureys, "Willful Modulation of Brain Activity and Communication in Disorders of Consciousness," *New England Journal of Medicine* 362 (2010): 579–89.

CHAPTER 10

Page 156 *A BBC film crew had asked if they could record the scanning session with Scott, which added, for me at least, an extra level of anxiety to it. The BBC had been following our work for their series* Panorama, *which was first broadcast in 1953 and is the world's longest-running current-affairs documentary program.*

The final award-winning documentary can be viewed online at www.intothegrayzone.com/mindreader.

Page 163 *I heard one story about a patient who loved the Canadian artist Celine Dion.*

Like many of the most interesting stories I have been told over the last few years, this tale was recounted to me by my brilliantly insightful graduate student Loretta Norton.

Page 164 *Scott told us otherwise. He answered all of those questions and more. When we asked him what year it was, he told us correctly that it was 2012, not 1999, the year of his accident.*

For further details, see D. Fernandez Espejo and A. M. Owen, "Detecting Awareness after Severe Brain Injury," *Nature Reviews Neuroscience* 14 (11) (2013): 801–9.

Page 165 *The most famous case of anterograde amnesia is Henry Molaison, or H.M., as he is more widely known.*

For further details, see W. B. Scoville and B. Milner, *Journal of Neurology, Neurosurgery and Psychiatry* 20 (1957): 11–21.

CHAPTER 11

Page 171 *"The healthy subject is taken with sudden pain; he immediately loses his speech and rattles his throat. His mouth gapes and if one calls him or stirs him he only groans but understands nothing."*

This quote is taken from the Hippocratic writings, translated by E. Clarke, "Apoplexy in the Hippocratic Writings," *Bulletin of the History of Medicine* 37 (1963): 307.

Page 171 *The term* pie vegetative *was used in 1963 and* vegetative survival *in 1971, predating the introduction of* persistent vegetative state *in a landmark paper published by Bryan Jennett and Fred Plum in the* Lancet *on April Fools' Day in 1972.*

Pie vegetative was used in a paper by M. Arnaud, R. Vigouroux, and M. Vigouroux, "États Frontieres entre la vie et la mort en neuro-traumatologie," *Neurochirurgia* (Stuttgart) 6 (1963): 1–21. *Vegetative survival* was used by M. Valpalahti and H. Troupp, "Prognosis for Patients with Severe Brain Injuries," *British Medical Journal* 3 (5771) (1971): 404–7. The classic paper by Bryan Jennett and Fred Plum was published as B. Jennett and F. Plum, "Persistent Vegetative State after Brain Damage: A Syndrome in Search of a Name," *Lancet* 299 (7753) (1972): 734–37.

Page 173 *At the Royal Society in London, I recently co-organized a meeting on consciousness and the brain with my friend and colleague Mel Goodale.*

The meeting was one of the first of the CIFAR Azrieli Program in Brain, Mind & Consciousness, which I codirect with my friend and colleague Mel Goodale. The meeting's title was "Biomarkers of Consciousness," and it was held at the Royal Society of London. Founded in 1660, the Royal Society has existed for centuries to "promote science and its benefits, recognise excellence in science, support outstanding science, and provide scientific advice for policy, foster international and global cooperation, education and public engagement." It's an institution in the grandest sense. To be elected a member of the Royal Society is among the highest accolades bestowed upon academics worldwide.

Page 173 *Think about boiling mussels. Relatively few people are troubled by throwing a bag of them into boiling water.*

I must thank my good friend and colleague of many years John Duncan of the MRC Cognition and Brain Sciences Unit, Cambridge, for initiating this lively debate. John was one of our guests at that meeting at the Royal Society, and I am fairly sure that he started the conversation with exactly this sentence.

Page 186 *Steven Laureys and his colleagues have conducted a study that suggests that what we all think we want to happen to us ("Please don't let me live in a gray-zone state") is not what we actually want when disaster strikes.*

In locked-in syndrome an individual is fully conscious, but unable to move or speak due to quadriplegia and anarthria. It is not generally considered to be a "disorder of consciousness," but is frequently confused as such. Where signs of awareness are missed (such as eye movements or eye blinking), locked-in patients can be mistaken for vegetative or minimally conscious. For further details, see M. A. Bruno, J. Bernheim, D. Ledoux, F. Pellas, A. Demertzi, and S. Laureys, "A Survey on Self-Assessed Well-Being in a Cohort of Chronic Locked-In Syndrome Patients: Happy Majority, Miserable Minority," *British Medical Journal Open*, 2011, 1:e000039, doi:10.1136/bmjopen-2010-000039.

CHAPTER 12

Page 191 *What Martin showed with his elegant experiment is that if you ask healthy participants in the fMRI scanner to focus first on the face and then on the house, activity in their brains would switch from the fusiform gyrus to the parahippocampal gyrus, exactly at the point that they made the attentional switch.*

For further details, see M. M. Monti, J. D. Pickard, and A. M. Owen, "Visual Cognition in Disorders of Consciousness: From V1 to Top-Down Attention," *Human Brain Mapping* 34 (6) (2012): 1245–53.

Page 194 *The idea came from a study that had been conducted almost ten years earlier by colleagues in Israel that had nothing to do with brain injury or disorders of consciousness.*

For further details, see U. Hasson, Y. Nir, I. Levy, G. Fuhrmann, and R. Malach, "Intersubject Synchronization of Cortical Activity during Natural Vision," *Science* 303 (2004): 1634–40.

Page 203 *In 1985, my Cambridge colleague Simon Baron-Cohen and his colleagues were the first to suggest that children with autism lack theory of mind.*

For further details, see S. Baron-Cohen, A. M. Leslie, and U. Frith, "Does the Autistic Child Have a 'Theory of Mind'?," *Cognition* 21 (1) (1985): 37–46.

Page 204 *When we published Jeff's story and our new approach to measuring consciousness in the prestigious journal the* Proceedings of the National Academy of Sciences *in 2014 . . .*

For further details, see L. Naci, R. Cusack, M. Anello, and A. M. Owen, "A Common Neural Code for Similar Conscious Experiences in Different Individuals," *Proceedings of the National Academy of Sciences* 111 (39) (2014): 14277–82.

CHAPTER 13

Page 220 *Back in my PhD days we'd shown that when you give Parkinson's patients a simple problem-solving task, they take far longer than healthy elderly people to find the solution, although they do get there in the end.*

For further details, see A. M. Owen, M. James, P. N. Leigh, B. A. Summers, C. D. Marsden, N. P. Quinn, K. W. Lange, and T. W. Robbins, "Fronto-striatal Cognitive Deficits at Different Stages of Parkinson's Disease," *Brain* 115 (pt. 6) (1992): 1727–51.

CHAPTER 14

Page 227 *At the same time, our explorations into the gray zone were helping us to unravel the building blocks of consciousness—how brain processes like memory, attention, and reasoning relate to unitary concepts like "intelligence" and how they emerge from that three-pound lump of gray and white matter inside our heads.*

In 2012, we published a scientific paper debunking the concept of "g" or general intelligence ("IQ") and described a new way for understanding differences in brain function in terms of memory, reasoning, and verbal abilities. We recruited more than 44,000 members of the general public and analyzed their performance on a wide variety of cognitive tests. If you would like to try this for yourself, the tests are available online at www.cambridgebrainsciences.com. For further details, see A. Hampshire, R. Highfield, B. Parkin, and A. M. Owen, "Fractioning Human Intelligence," *Neuron* 76 (6) (2012): 1225–37.

Page 237 *What was being played through the headphones was a cornucopia of words and phrases dreamed up by Damian, carefully designed to unearth what might be going on in Leonard's brain.*

For further details, see S. Beukema, L. E. Gonzalez-Lara,

P. Finoia, E. Kamau, J. Allanson, S. Chennu, R. M. Gibson, J. D. Pickard, A. M. Owen, and D. Cruse, "A Hierarchy of Event-Related Potential Markers of Auditory Processing in Disorders of Consciousness," *NeuroImage Clinical* 12 (2016): 359–71.

CHAPTER 15

Page 244 *Axel and Tim, together with their colleague Patrick Wilken, have produced the excellent* Oxford Companion to Consciousness.

For more details, see T. Bayne, A. Cleeremans, and P. Wilken, *The Oxford Companion to Consciousness* (New York: Oxford University Press, 2009).

Page 246 *One system presents people with a screen display of letters from* A *to* Z *and asks them to focus their attention on specific letters.*

For further details, see L. A. Farwell and E. Donchin, "Talking off the Top of Your Head: Toward a Mental Prosthesis Utilizing Event-Related Brain Potentials," *Electroencephalography and Clinical Neurophysiology* 70 (1988): 510–23.

Page 247 *One way around this is to place the electrodes directly on the brain's surface—a complex neurosurgical procedure for sure, but one that can produce miraculous results.*

For more details, see L. R. Hochberg, D. Bacher, B. Jarosiewicz, N. Y. Masse, J. D. Simeral, J. Vogel, S. Haddadin, J. Liu, S. S. Cash, P. van der Smagt, and J. P. Donoghue, "Reach and Grasp by People with Tetraplegia Using a Neurally Controlled Robotic Arm," *Nature* 485 (2012): 372–75.

Page 249 *A recent episode of the TNT show* Perception *used our research to build a plotline involving almost exactly this scenario.*

Perception, Season 1, Episode 4, "Cipher," directed by Deran Serafian, written by Jerry Shandy (TNT, 2012). The relevant scene can be viewed at www.intothegrayzone.com/perception.

Page 250 *The recent documentary* Alive Inside, *which won the Audience Award at the 2014 Sundance Film Festival, chronicles the astonishing experiences of several people with Alzheimer's disease whose lives were turned around when they were played music that they had known and loved.*

Alive Inside: A Story of Music and Memory, written and directed by Michael Rossato-Bennett, produced by Projector Media and the Shelley & Donald Rubin Foundation (2014).

Page 254 *Collective consciousness has features in common with what some have termed universal or cosmic consciousness. Universal consciousness is said to be "an infinite, eternal ocean of intelligent energy . . ."*

I found this relatively recent description of "universal consciousness" at www.loveorabove.com/blog/universal-consciousness. The term "cosmic consciousness" was coined in 1901 by the Canadian psychiatrist Richard Maurice Bucke in his book *Cosmic Consciousness: A Study in the Evolution of the Human Mind* (Philadelphia: Innes & Sons, 1901).

INDEX

INDEX

INDEX

What has surprised you most about your journeys into the gray zone?

Working with people in the gray zone—and by that, I mean patients who hover at the border between life and death—I've been surprised by many things over the years, but perhaps most obviously that the relationship between what people appear to be and what they actually are is far less clear than we used to think it was. For many years, patients who would not respond to any form of external stimulation, like a request to move a hand or blink an eye, were presumed to be in a so-called vegetative state—"awake, but unaware." When we showed in 2006 that fMRI could be used to reveal covert awareness in some of these patients, it became immediately apparent what a grave error in judgment that had been.

So what do we now know?

That many of these patients harbor some level of awareness; indeed, some of them are completely conscious, locked inside their own bodies, yet unable to communicate with the outside world in any conventional way. Moving on to start communicating with some of those patients in 2010, again using fMRI, was another major milestone for us. I guess it was less of a "surprise" at that stage, because by then we knew the patients were in there; it was just a matter of developing the technology to the point that we could open up a reliable channel of communication with them. But it was a thrilling achievement nonetheless. If I were to get a little more philosophical about what has surprised me most about working in this area, I would have to say that it has convinced me more and more that we really are our brains. As I wrote in *Into the Gray Zone* [www.intothegrayzone.com], "[Y]our brain is who you are. It's every plan you've ever made, every person you've fallen in love with, and every regret you've ever had. Your brain is all there is. Without a brain, our sense of 'self' is reduced to nothing."

Not all psychologists would agree with that.
I know. Dick Passingham, who is a great friend of mine and someone I have respected for many years, recently pulled me up on this. . . . He pointed out that there is a whole lot more to us than that. As well as our brains, we are our bodies and our past histories, he said. I replied that we can replace almost any part of our bodies, and we will remain the exact same person. But swap out someone's brain and there is no question that they will cease to be them; their memories, their personality, their attitudes and beliefs, everything will change, including their entire past history (at least as they recall it). The very essence of them as a person will be forever altered.

Do you have any estimate of how many of these patients might be out there, awaiting that "magic," to be "found"?
Improvements in roadside medicine and intensive care have led to more people surviving serious brain damage and ending up alive but with no evidence of preserved awareness. We don't know exactly how many vegetative-state patients there are, nor how many of them are trapped in the "gray zone," awake and aware, but unable to convey that fact to the outside world. This is due, in large part, to poor or noncentralized nursing-home records. In the United States, estimates range from 15,000 to 40,000 vegetative-state patients. Given that number, our findings suggest that as many as 7,000 of these patients might actually be aware of everything

ABOUT THE AUTHOR

Adrian Owen is the Canada Excellence Research Chair (CERC) in Cognitive Neuroscience and Imaging at the University of Western Ontario, Canada, and has spent the last twenty-five years pioneering breakthroughs in brain science. His research combines structural and functional neuroimaging with neuropsychological studies of brain-injured patients and has been published in many of the world's leading scientific journals, including *Science*, *Nature*, the *New England Journal of Medicine*, and the *Lancet*. His findings have attracted widespread media attention on TV and radio, in print, and online and have been the subject of several TV and radio documentaries. Since 1990, Professor Owen has published more than three hundred articles and chapters in scientific journals and books.

ADRIAN OWEN

AT THE FLIMSY BORDER BETWEEN LIFE AND DEATH

Jon Sutton Meets Adrian Owen
to Discuss His Book, *Into the Gray Zone*

going on around them. Such patients can be found in virtually every city and town in the world with a skilled nursing facility.

A while back I sent you my own draft "Advance Decision," and while saying you do advocate them, you cautioned, "Imagine again that you are aware and have to witness your own death just because you used to think that was a good idea." How do you deal with that conundrum?

Great question! The reason I said that back then was because I was beginning to realize that what most of us think we would want if we are unfortunate enough to end up in the gray zone, is not actually what we want when we get there. How many times have you heard (or said) "I don't want to end up like that—pull the plug!" Through technology like fMRI, we have been able to communicate with some of these patients, and we now know that their lives are not necessarily as bleak as we might think they are. That's not to say that they are blissfully happy, but consider this example: My colleague Steve Laureys recently surveyed ninety-one people with locked-in syndrome—conscious people who were only able to communicate by blinking or vertically moving their eyes. They were asked to answer a questionnaire about their medical history, current status, and attitude toward end-of-life issues. Contrary to what most of us might expect, a significant proportion of the patients (72 percent of those who responded) reported that they were happy. What's more, a longer time in a locked-in syndrome was correlated with how happy this group said they were.

Not everyone has been convinced by your work, pointing to physiological and statistical artifacts and cautioning against overinterpretation. How do you respond?

This is a great question that gives me an opportunity to point out that I don't get to spend my whole life just doing cool science! In fact, I spend a lot of time managing people's expectations, juggling the demands and interests of doctors, patients, and families, and justifying the scientific conclusions we have reached to people who may not believe, or understand, what we have discovered. For me, this is the "politics of science," and it's a necessary evil. We can't just sit in our ivory towers discovering stuff, writing about it in scientific journals and hoping that everyone will just take it at face value. I've generally found that it's very worthwhile taking the time to explain our findings, sometimes even running additional experiments to rule out artifacts or potential design flaws, and usually the end result is a better understanding of the data and a more accurate interpretation of the results.

In many instances, it's been the criticisms that have driven me on toward the next big question. For example, I remember after the 2006 paper came out in *Science*, a small but vocal contingent doubted our findings; they said that fMRI activity alone was not sufficient to conclude that our patient was conscious. This skepticism led directly to our next step—to actually communicate with a patient who was presumed to be vegetative. My reasoning at the time was, if they don't believe the patient is conscious, what better way than to get the patient to tell us themselves! But that said, overall, I have had tremendous support from both the scientific community and the public for my work, and I am very grateful for that.

Very few psychologists—sorry, "neuroscientists"!—can claim to have pioneered a new field of science. Do you put this down to vision, luck, or a bit of both?

I would never claim to have pioneered a new field of science—apart from anything else. I have worked with amazing students, postdocs, and collaborators for many years, and they have all contributed in great and small ways to the emergence of this field. In many ways, my role has been to bring these people together—often from disciplines that might never normally talk to each other, like engineers, ethicists, psychologists, and critical care staff. That's when the magic really happens!

But I have also been very lucky; I trained in psychology in the late 1980s, and that really equipped me well for the emergence of cognitive neuroscience and brain imaging. When PET and then fMRI came along in the early 1990s, I think I was really well positioned to get into those fields using sound psychological principles. In those early days that was really important, because there was a lot of really terrible brain imaging being done in the name of psychology! Then, having that solid grounding in brain imaging allowed me to look beyond simply "brain mapping," which is what I did for most of the 1990s, and find a real application for the technology in a clinical setting. I wouldn't say I ever had a "vision" for any of this, but perhaps it helped to have a sense of knowing what I knew, and the foresight to try to apply that knowledge to novel situations and scientific questions.

So you're a psychologist—one of the last (as in, youngest) of the "old guard" of British psychologists, the UCL/APU school. Familiar figures crop up in the book, such as John Duncan, Pat Rabbitt, and Trevor Robbins. Can you understand us having a little pang of jealousy at seeing "A neuroscientist explores . . ." on the cover?

I am very proud to have trained in psychology, and I think it provided me with many of the tools that I have used over the years to tackle the scientific questions that have interested me, from how to design a rigorous experiment to an appreciation of medical ethics. But there came a time when describing myself as a "psychologist" no longer told people much about what I actually do. In fact, it often gave them the wrong impression entirely. One of the themes running through *Into the Gray Zone* is how the field that we now know as "cognitive neuroscience" emerged out of psychology with a lot of help from brain imaging, and my career followed that very same path across the same time span. I am a psychologist at heart, but a lot of what I do is better described as neuroscience. I like to think that what I am best at is applying good psychological principles to neuroscientific questions, some of which may not seem very "psychological" at all.

You and your teams have had so many successes. But, like neurosurgeon Henry Marsh in *Do No Harm,* you seem somewhat inevitably haunted by the failures. Peering into your past girlfriend Maureen's brain, only to see "darkness. A void," or being bold enough to ask John "Do you want to die?" and getting inconclusive results.

I don't think I am haunted by the failures, but ups and downs are the reality of science, particularly when you are pursuing a single, burning question over many years. The problem is, in the scientific papers, readers often only get to see the highlights . . . the net result of several experiments and the summary of what we think it all means. I'm not talking about the tendency not to publish negative results—that's a different type of problem—but rather, all the ideas that never got off the ground, all the unfortunate twists and turns that got in the way of the science. In our case, patients dying (or recovering) just as we're about to make contact with them is a fairly common occurrence. Of course, our first thoughts are with the families, but I can't pretend that these things aren't also scientifically disappointing because often it sends us right back to the drawing board. Writing the book gave me the space and the freedom to write about these things and express how they make us, as scientists, feel. Of course, the science is in there in a way that I hope everyone will be able to understand, but there's also a lot about the process of doing science, the ups and the downs, the triumphs and the failures, the setbacks and the sense of sheer exhilaration that I still get when we do occasionally make a breakthrough. What I really wanted to convey in *Into the Gray Zone* is what it's actually like to be a scientist, set against a backdrop of the scientific discoveries themselves and, of course, the incredible stories of the patients and their families.

That really does come across in the book, which is beautifully written. Where did that come from? . . . Did you find it easy? Did it surprise you?
I resisted requests to write *Into the Gray Zone* for many years, because I wasn't really sure that I could do justice to the patients and their extraordinary stories. I have written hundreds of scientific papers, but this required a different kind of approach. Fortunately, our research has received plenty of attention from the media and the public over the years, and this has given me a reasonably good sense of how to write for (and talk to) the nonexpert, and that made it much easier than I expected.

One sidenote that might be of interest to your readers: I began writing these stories against a backdrop of the relevant neuroscience, focusing on the anatomy of the brain, the function of various regions, and so on, but quite quickly realized that wasn't where my heart was and that wasn't going to be a very easy book to write. About two chapters in, I realized that it was the psychological aspects of the scientific journey that fascinated me most, why we think the way we think and why that is such a critical and vulnerable part of the human condition. Once I'd found that voice, it became much easier to keep writing.

You dedicate the book to your son, saying "In case I'm not here to tell you the story myself." That struck me as a little fatalistic! Then I realized you had returned to that point in the acknowledgments . . .

I'm glad you noticed that. It was always my intention to dedicate the book to my son, Jackson, because it was written during the first three years of his life, in between (and sometimes during!) nappy changes, midnight feedings, and many, many sleepless nights. The story also covers a period of my life that in many ways defines who I am now as a person. When I read a book dedication, I always feel a bit shortchanged when there's no explanation. "For Jackson" doesn't tell you anything or why this book is for Jackson. I added "In case I'm not here to tell you the story myself" because I hope that one day he'll read it and understand something about my life that won't be obvious from the daily interactions he and I share as father and son. I hope I'll still be around to add additional color, of course, but something I have learned from working in this area is that there is absolutely no way to predict that. As I wrote at the end of the book, "working at the flimsy border between life and death, the sheer fragility of every human life is hard to ignore."

Have you found a growing need for some "light relief" in other areas of your personal and professional life?

I have never been a terribly serious person, and I have always pursued things that interest and entertain me. But it's tough to try to fit it all in. One way is to just combine the two. I play guitar and sing in a band called Untidy Naked Dilemma, made up entirely of neuroscientists, and that's been an incredibly important part of my life for many years. In various forms, the bands I've played in have performed all over the world, often at academic conferences that many of the members of the band were attending.

From time to time, I'll also do a fun scientific study with someone who is an academic friend rather than a regular collaborator, just because we can and because it's fun. Richard Wiseman and I, who have been close friends for more than thirty years now—we were psychology undergraduates together at UCL between 1985 and 1988—recently published a study together pointing out that about 80 percent of the time, pictures of the human brain are presented facing to their right (that is, we mostly get to see the left hemisphere). Why that is, we do not know, but it's one of those quirky little things that Richard spotted, and we decided to write an academic paper about it. It's wonderful to have the freedom to do things like that with your friends, especially after thirty years in the field.

In any area like this, dark humor can be a defense mechanism, and there's a bit of that in the book. The locked-in patient subjected to one Celine Dion album springs to mind. What would be your own torture track—something by your own band, maybe?

If I were ever unfortunate enough to find myself in the gray zone, then I am fairly sure that my students would put me straight into the fMRI scanner and ask me to imagine playing tennis. Although it might well help them work out whether I'm in there or not, after all this time saying those words, I'm not really sure I ever want to hear them said to me!

There is quite a bit of humor in the book, and that was always intentional. Despite some of the awful things we have to witness on a day-to-day basis, funny things do happen, and I included them in the story because that's the reality of the situation, and I wanted to remain as honest as I could to that reality. I can imagine that people might think it's terribly depressing working closely with people at the border between life and death, but there are plenty of funny moments, surprising moments, joyous moments, and exhilarating moments, and I really wanted that to come through in the pages of the book. When you're working with people who are living with terrible life circumstances, it's important to keep some perspective, or it really would be very hard to go home and sleep at night.

Oh, and it would have to be "The Final Countdown" by Europe, mostly because it haunted my undergraduate years at UCL (it was number one on the UK charts in 1986), but also because the cruel irony would be simply too much to live with . . .